STUDENT ACTIVITY BOOK

to accompany

MATHEMATICS FOR ELEMENTARY SCHOOL TEACHERS

by Richard J. Sgroi *and* Laura Shannon Sgroi

Victoria J. Garrison

State University of New York/New Paltz

PWS Publishing Company/Boston

PWS
Publishing Company

20 Park Plaza
Boston, Massachusetts 02116

PWS Publishing Company is a division of Wadsworth, Inc.

ISBN: 0-534-93258-4

Printer: Malloy Lithographing

Printed in the United States of America.
2   3   4   5   6   7   8   9   --   98   97   96   95

PREFACE

This *Student Activity Book* helps you develop a better understanding of the concepts presented in MATHEMATICS FOR ELEMENTARY SCHOOL TEACHERS by Richard J. Sgroi and Laura Shannon Sgroi through supplemental activities relating to each chapter of the main text.

These activities make use of many of the traditional manipulatives found in elementary school classrooms. Several of the more useful manipulatives are provided in the back of this book and others are found with the main text. Each activity instructs you as to which of the manipulatives you will need to successfully complete the work required. Activity numbers are coordinated with the chapter and section numbers in MATHEMATICS FOR ELEMENTARY SCHOOL TEACHERS. There is an activity for each major idea presented in the main text.

Each chapter opens with an activity that relates the chapter topic to your own day-to-day experiences. This is to promote an increased appreciation of the variety of mathematical applications you encounter daily and of which you may be unaware.

"Check Your Understanding" pages are included for you to monitor your progress. Your instructor may wish to use these in the classroom, and for that purpose, the answers are provided in the *Instructor's Manual*.

Answers to each of the activities, many of which include detailed explanations, are provided in the back of this book.

It is my hope that these activities will facilitate your understanding of the concepts you encounter in MATHEMATICS FOR ELEMENTARY SCHOOL TEACHERS and provide you with the confidence needed in your future teaching of elementary mathematics.

Victoria J. Garrison

ACKNOWLEDGMENTS

There are many individuals whose support during the development of this project is greatly appreciated. Special thanks go to the following:

Richard J. Sgroi and Laura Shannon Sgroi for their enthusiasm and their confidence in me.

Barbara Lovenvirth of PWS Publishing for her patient cooperation when the distance between us increased by approximately 5000 miles.

Roland Bowen of Santiago, Chile for the use of his laser printer whenever it was requested of him.

My friends and family, especially my colleague, Dr. Salvatore Anastasio, and my parents, Howard and Phyllis Garrison, for their constant encouragement and their interest in my work.

V.J.G.

CONTENTS

MANIPULATIVES

Chapter 1:

    Pattern blocks from Appendix

    Attribute blocks from Appendix

Chapter 2:

    Chip trading board from Appendix

    Red, white, & blue chips from Appendix

    Base three pieces from Appendix

    Base ten pieces from the Appendix

Chapter 3:

    Centimeter rods included with textbook

    Base ten pieces from Appendix

    Base three pieces from Appendix

Chapter 4:

    Centimeter rods included with textbook

    Square tiles from Appendix

    Attribute blocks from Appendix

    Function machine visuals from Appendix

    Clock 12 card from Appendix

    Modulo three card from Appendix

Chapter 5:

    Chips & box diagram from Appendix

Chapter 6:

    Fraction circles from Appendix

    Tangrams from Appendix

    Centimeter rods included with textbook

Chapter 7:

Base ten pieces from the Appendix

Chapter 8:

Metric tape measure from the Appendix

Chapter 9:

Attribute blocks from Appendix

Chapter 10:

Polygon cards from Appendix

Tangrams from Appendix

Geoboard included with textbook

Models of prisms and pyramids from Appendix

Chapter 11:

Protractor from Appendix

Geoboard included with textbook

Square tiles from Appendix

Surface area models from Appendix

Pyramid and prism models from Appendix

"Volume Models" from Appendix

Chapter 12:

NONE

Chapter 13:

NONE

1: MATHEMATICAL CONNECTIONS

Introduction: The visual, verbal, and symbolic patterns in our world facilitate prediction and problem solving. The activities in this section will encourage your awareness of these patterns.

1. Find three decorative patterns in magazines, wallpaper, fabric, etc. Attach copies of them in the spaces provided below. Explain the pattern.

    a.

    b.

    c.

2. Name a pair of common objects that are the same size and shape. Objects that have the same size and shape are congruent. How can you demonstrate the congruency of the objects that you have named?

3. Describe how patterns aid in prediction in each of the following areas.

    a. Weather forecasts

    b. Libraries

    c. Music

ACTIVITY 1.1A - PATTERNS & CONGRUENCY

Materials required: *Tracing paper, patterns you chose for page 1*

1. Using tracing paper, predict the next section of each of the patterns that you used on the previous page. Attach the tracing paper in the space provided next to each letter.

    a.

    b.

    c.

2. Identify the number of pairs of congruent triangles in each of the figures below.

a.

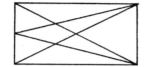

b.

3. Identify the number of congruent figures in each of the diagrams below.

a.

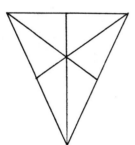

b.

ACTIVITY 1.1B - SYMMETRY

Materials required: Pattern blocks from the Appendix, scissors

1. Instructions: Cut out the pattern blocks found at the back of this book. Examine each of the figures for horizontal, vertical, diagonal, and rotational symmetry.

PATTERN BLOCK #	HORIZONTAL?	VERTICAL?	DIAGONAL?	ROTATIONAL?
1				
2				
3				
4				
5				
6				

2. Using at least two different pattern blocks #1 − #6, create patterns which exhibit the following types of symmetry. Trace the patterns in the space below.

   a. Rotational symmetry

b. Diagonal line symmetry

c. Horizontal line symmetry

d. Vertical line symmetry

1.1 CHECK YOUR UNDERSTANDING

1. Describe one way in which to determine if two figures are congruent.

2. Give an example of a figure which is asymmetric.

3. Determine the types of symmetry displayed by each of the following figures.

FIGURE	HORIZONTAL	VERTICAL	DIAGONAL	ROTATIONAL

ACTIVITY 1.2A - VENN DIAGRAMS AND SET OPERATIONS

Materials *required*: *Three loops of string, attribute blocks from the Appendix*

1. Let the universe consist of all attribute blocks. Position the loops of string and attribute pieces to represent the following situations. Sketch the result in the space provided at the right.

  a. Let A = {yellow blocks}
     Let B = {small circles}
     Let C = {small blocks}

  b. Let A = {blue circles}
     Let B = {blue blocks}
     Let C = {blue triangles}

  c. Let A = {large triangles}
     Let B = {blue blocks}
     Let C = {blue triangles}

  d. Let A = {all squares}
     Let B = {small squares}
     Let C = {red blocks}

2. Use your attribute blocks to determine A − B for the situations in #1 on the previous page. Record your results below.

   a.

   b.

   c.

   d.

3. Use your attribute blocks to determine B − A for each situation in #1 on the previous page. Record the results below.

   a.

   b.

   c.

   d.

4. Is determining the difference of two sets a commutative operation? Explain using #2 and #3 above.

5. Use your attribute blocks to determine A ∪ (B ∩ C) for each of the situations in #1. List the elements below.

   a.

   b.

   c.

   d.

6. Use your attribute blocks to determine (A ∪ B) ∩ C for each of the situations in #1. List the elements below.

   a.

   b.

   c.

   d.

7. Is A ∪ (B ∩ C) equivalent to (A ∪ B) ∩ C? Explain your answer.

8. Use your attribute blocks to determine A ∩ (B ∩ C) for each situation in #1. List the elements below.

   a.

   b.

   c.

   d.

9. Describe the elements of each set in #8 a - d in your own words.

    a.

    b.

    c.

    d.

10. Is $A \cap (B \cap C)$ equivalent to $(A \cap B) \cap C$? Explain.

11. Use your attribute blocks to determine $A \cup (B \cup C)$ for each situation in #1. List the elements below.

    a.

    b.

    c.

    d.

12. Describe the elements of each set in #11 a - d in your own words.

    a.

    b.

    c.

    d.

13. Is $A \cup (B \cup C)$ equivalent to $(A \cup B) \cup C$? Explain.

ACTIVITY 1.2B - THE COMPLEMENT OF A SET

Materials required: Two loops of string, attribute blocks from the Appendix

1. Let your universe be the set of triangles. Set A consists of all small triangles and set B consists of all red triangles. Represent the relationship between sets A and B in the diagram below. Show where each block belongs.

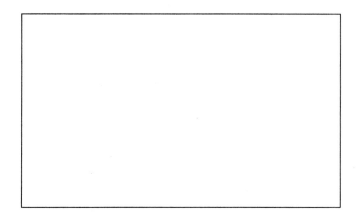

Represent each of the following using your attribute blocks. List the members of each set.

a. A ∩ B

d. A ∪ B

b. $\overline{A \cap B}$

e. $\overline{A \cup B}$

c. $\overline{A} \cap \overline{B}$

f. $\overline{A} \cup \overline{B}$

2. Which parts of #1 are equivalent? List them.

3. Let your universe be the set of yellow attribute blocks. Set A consists of all large circles and set B consists of all squares. Represent the relationship between sets A and B in the diagram below. Show where each block belongs.

Represent each of the following using your attribute blocks. List the members of each set.

a.   A ∩ B

d.   A ∪ B

b. $\overline{A \cap B}$

e. $\overline{A \cup B}$

c. $\overline{A} \cap \overline{B}$

f. $\overline{A} \cup \overline{B}$

4. Which parts of #3 are equivalent? List them.

5. What conclusions can be drawn from #2 and #4?

ACTIVITY 1.2C - VENN DIAGRAMS AND SETS

Materials required: Attribute blocks from the Appendix

INSTRUCTIONS: Let the universe consist of all attribute blocks.   For each diagram describe a possible assignment for sets A, B, and C.

1.

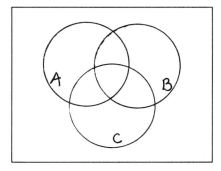

A =

B =

C =

2.

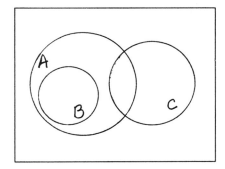

A =

B =

C =

3.

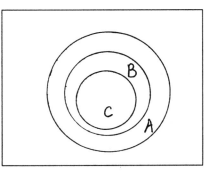

A =

B =

C =

4.

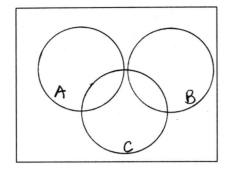

A =

B =

C =

5.

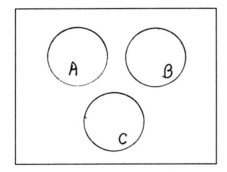

A =

B =

C =

6.

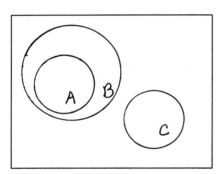

A =

B =

C =

7.

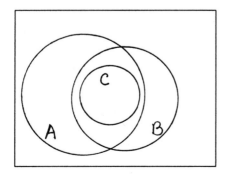

A =

B =

C =

16

1.2 CHECK YOUR UNDERSTANDING

Let the universe be your set of attribute blocks and A = {red blocks} , B = {small blocks}, and C = {all circles).

1. Draw a Venn diagram to represent the relationship among sets A, B, and C.

2. List the elements in each of the following.

    a.   A ∪ (B ∩ C)

    b.   A ∩ (C − B)

3. Does $A - B = B - A$? Explain.

4. Does $\overline{A \cup B} = \overline{A} \cup \overline{B}$? Explain.

5. Describe $A \cap (B \cap C)$ in words.

1.3A - NEGATION, CONJUNCTION, & DISJUNCTION

Let p: The block is blue.
q: The block is a triangle.

I. Consider the truth value of the statements above using each of the blocks given below.

	p	q
1. large blue triangle		
2. small blue circle		
3. small red triangle		
4. large red circle		

II. <u>Instructions</u>: Express each of the following in words and complete each chart.

A.   NEGATION.   ~p: _____

~q: _____

	p	q	~p	~q
1. large blue triangle				
2. small blue circle				
3. small red triangle				
4. large red circle				

SUMMARY: ~p is false when _____

~q is false when _____

19

B. CONJUNCTION. **p ∧ q** : _

This means the block must be <u>BOTH</u> _ _ _ _ _ _ _ _ _ _ AND _ _ _ _ _ _ _ _ _ _ .

	p	q	p ∧ q
1. large blue triangle			
2. small blue circle			
3. small red triangle			
4. large red circle			

SUMMARY: **p ∧ q** is false when _

C. DISJUNCTION. **p ∨ q** : _

This means the block may be _ _ _ _ _ _ _ _ _ OR _ _ _ _ _ _ _ _ _ OR <u>BOTH</u> _ _ _ _ _ _ _ _
AND _ _ _ _ _ _ _ _ _ _ .

	p	q	p ∨ q
1. large blue triangle			
2. small blue circle			
3. small red triangle			
4. large red circle			

SUMMARY: **p ∨ q** is false when _

Let p: The block is blue.
q: The block is a triangle.

I. CONDITIONAL.

A. Express $p \rightarrow q$ in words. _____

B. If the statement in A is assumed to be a true statement, then whenever we have a piece that is

_____ it must also be _____. Otherwise we will contradict this

assumption and $p \rightarrow q$ will be false.

C. Complete the following table.

	p	q	$p \rightarrow q$
			Is the assumption contradicted? If YES, write **F.** If NO, write **T.**
1. large blue triangle			
2. small blue circle			
3. small red triangle			
4. large red circle			

D. $p \rightarrow q$ is false when _____

II. BICONDITIONAL.

A. Express $p \leftrightarrow q$ in words. _____

This means $p \rightarrow q$ AND $q \rightarrow p$. In words, this means _____

_____.

B. If the statement in A is assumed to be true, then whenever we have a piece that is _____

it must also be _____ AND whenever we have a piece that is _____

it must also be _____. Otherwise the assumption will be contradicted and

$p \leftrightarrow q$ will be false.

C. Complete the following table.

	p	q	p → q	q → p	$p \leftrightarrow q$ Is the assumption contradicted? If YES, write **F.** If NO, write **T.**
1. large blue triangle					
2. small blue circle					
3. small red triangle					
4. large red circle					

D. $p \leftrightarrow q$ is false when _____.

1.3C - CONVERSE, INVERSE, & CONTRAPOSITIVE

Let p : The number is greater than seven.
q : The number is greater than four.

I. Write each of the following in words.

   a. Statement: $p \rightarrow q$

   b. Converse: $q \rightarrow p$

   c. Inverse: ${\sim}p \rightarrow {\sim}q$

   d. Contrapositive: ${\sim}q \rightarrow {\sim}p$

II. Give the truth value of each of the statements in parts a - d above. If a statement is false, then explain.

   a.

   b.

   c.

   d.

III. Which of the following have the same truth values?

      STATEMENT   CONVERSE   INVERSE   CONTRAPOSITIVE

23

1.3D - LAWS OF LOGIC

Let p : The number is divisible by twelve.
q : The number is divisible by four.
r : The number is divisible by two.

I. LAW OF DETACHMENT (MODUS PONENS): $[(p \rightarrow q) \land p] \rightarrow q$

p → q :   If the number is _____, then the number is _____.

p :   We know the number is _____.

Therefore **q :**   The number is _____.

Complete the following table.

p	q	p → q	(p → q) ∧ p	[(p → q) ∧ p] → q

When is this argument false? Is it a tautology? Explain.

II. LAW OF MODUS TOLLENS: $[(p \rightarrow q) \wedge \sim q] \rightarrow \sim p$

$p \rightarrow q$:   If the number is _____, then the number is _____.

$\sim q$:   We know the number is _____.

Therefore $\sim p$:   The number is _____.

Complete the following table.

p	q	~p	~q	p → q	(p → q) ∧ ~q	[(p → q) ∧ ~q] → ~p

When is the argument false? Is it a tautology? Explain.

III. LAW OF SYLLOGISM. $[\,(\,p \to q\,)\,\wedge\,(\,q \to r\,)\,]\,\to\,(\,p \to r\,)$

$\mathbf{p \to q}$: If the number is _____, then the number is _____.

$\mathbf{q \to r}$: If the number is _____, then the number is _____.

Therefore $\mathbf{p \to r}$: If the number is _____, then the number is _____.

Complete the following table.

p	q	r	p→q	q→r	p→r	(p→q)∧(q→r)	[(p→q)∧(q→r)] → (p→r)

When is this argument false? Is it a tautology? Explain.

1.3 CHECK YOUR UNDERSTANDING

1. Construct a truth table for each of the following statements.

    a. $(p \rightarrow q) \wedge \sim p$

    b. $(p \vee q) \leftrightarrow \sim q$

2. Write the converse of the following statement.

        If it is raining, then I do not have my umbrella.

3. Is the statement given in #2 logically equivalent to its converse? Explain.

4. Let p: I am in love.

    q: I am singing.

    r: You are covering your ears.

    Write each of the following arguments in words.

    a. $[(p \rightarrow q) \land p] \rightarrow q$

    Hypotheses: _____

    _____

    Conclusion: _____

    b. $[(p \rightarrow q) \land \sim q] \rightarrow \sim p$

    Hypotheses: _____

    _____

    Conclusion: _____

    c. $[(p \rightarrow q) \land (q \rightarrow r)] \rightarrow (p \rightarrow r)$

    Hypotheses: _____

    _____

    Conclusion: _____

2: NUMERATION AND PLACE VALUE

Introduction: Activities in this section address the basic structure of our numeration system. A knowledge of this structure will enhance our recognition of the patterns which govern operations within the system.

History and Operation of the Abacus

I. Research the history of the abacus. Who used it? Where was it used? When was it used?
   Is it still used today?

II. Describe the way in which one uses an abacus. How does it operate? Are place value and grouping involved? Is there a basic principle of operation common to all of them?

ACTIVITY 2.1A - CREATE A NUMERATION SYSTEM

1. Design your own numeration system using any base you choose. Make a table of the basic symbols of your system and the Hindu-Arabic equivalent of each symbol. Include the symbols which will be required to represent numbers less than or equal to ten thousand.

SYMBOL	HINDU-ARABIC EQUIVALENT

2. Explain the design of your system.

   a. Is it additive?

   b. Is it multiplicative?

   c. Does it use a generalized grouping scheme?

   d. Does it use a base?

   e. Does use place value?

   f. Does it have a zero?

3. Represent each of the following numbers in your system.

    a. 4

    b. 27

    c. 319

    d. 6008

4. Which operations are easily performed in your system? Which are difficult? Explain.

    a. Addition

    b. Subtraction

    c. Multiplication

    d. Division

ACTIVITY 2.1B - PLACE VALUE WITH CHIP TRADING

Materials required: *Chip trading board, die, red chips, blue chips, white chips from the Appendix*

I. Using Base Three

  A. Roll the die.

  B. Place blue chips in the blue column equal in number to the dots on the die.

  C. Trade each group of three blue chips for a white chip in the white column.

  D. Trade each group of three white chips for a red chip in the red column.

  E. Record the result in the table below.

  F. Repeat four times (for a total of five times).

	R	W	B
1.			
2.			
3.			
4.			
5.			

  G. When trading is completed, what is the maximum number of chips which may appear in any of the columns?

  H. The blue column represents groups of _ _ _ _ _ _ _ .

   The white column represents groups of _ _ _ _ _ _ _ .

   The red column represents groups of _ _ _ _ _ _ _ _ .

I. If we add a green column at the left of the red column it would represent groups of

_____ .

J. What does each of the following represent?

    1. 211_{three}

    2. 120_{three}

II. Using Base Six
   A. Roll the die.
   B. Place blue chips in the blue column equal in number to the dots on the die.
   C. Trade each group of six blue chips for a white chip in the white column.
   D. Trade each group of six white chips for a red chip in the red column.
   E. Record the result in the table below.
   F. Repeat nine times (for a total of ten times).

	R	W	B
1.			
2.			
3.			
4.			
5.			
6.			
7.			
8.			
9.			
10.			

III. Explain what each of the following represents. What is the base ten equivalent of each?

   A. 241_{seven} B. 23_{nine} C. 2000_{five}

ACTIVITY 2.1C - PLACE VALUE WITH BASE THREE PIECES

Materials required: Base three pieces from the Appendix, die

I. The Base Three Pieces

    A. Value

        1. The blue pieces are units.

        2. The white pieces are longs.

        3. The red pieces are flats.

        4. The green piece is a block.

    B. Relationships

        1. How many units are in a long? _ _ _ _ _ .

        2. How many flats are in a block? _ _ _ _ _ .

        3. How many units are in a flat? _ _ _ _ _ .

        4. How many longs are in a flat? _ _ _ _ _ .

    C. Represent 211_{three} as two flats one long one unit. How would you represent 102_{three} ?

II. Chip Trading

    A. Roll the die.

    B. Make a pile of units equal in number to that on the die.

    C. Trade each group of three units for a long.

    D. Trade each group of three longs for a flat.

    E. Trade each group of three flats for a block.

    F. Record the result in the table below.

    G. Repeat five times. (For a total of six.)

Trial	B	F	L	U
1				
2				
3				
4				
5				
6				

ACTIVITY 2.1D - PLACE VALUE WITH BASE TEN PIECES

Required materials: Base ten pieces from the Appendix

I. Base Ten Pieces

   A. Value

      1. The blue pieces represent units.

      2. The white pieces represent longs.

      3. The red pieces represent flats.

      4. The green piece represents a block.

   B. Relationships

      1. How many units are in a long?   _____

      2. How many longs are in a flat?   _____

      3. How many units are in a flat?   _____

      4. How many units are in a block?   _____

      5. How many longs are in a block?   _____

      6. How many flats are in a block?   _____

II. Represent each of the following numbers using the base ten pieces.

	B	F	L	U
A. 235				
B. 402				
C. 1006				

III. How would you explain the comparison of 530 and 503 to a child using your base ten pieces?

2.1 CHECK YOUR UNDERSTANDING

1. In base seven we use only the digits _____.

2. What are the first four whole number place values (from right to left from the decimal point) in base four?

3. In base six you are given the following collection. Make the appropriate trades and represent the number in both <u>base</u> <u>six</u> <u>and</u> <u>base</u> <u>ten</u>.

    8 flats   5 longs   10 units

4. Explain what is represented by 1021_{three}. On the back of this page make a sketch which illustrates the meaning. Then give the base ten representation.

ACTIVITY 2.2A - USING A CALCULATOR

Materials *required*:   *Calculator*

I. Constant Arithmetic Feature

Enter each of the following keystroke sequences and give the result. Explain each operation.

A. $\boxed{\text{A C}}$ 5 $\boxed{+}$ $\boxed{=}$ $\boxed{=}$ $\boxed{=}$ $\boxed{=}$ Result:

Explain:

B. $\boxed{\text{A C}}$ 3 $\boxed{+}$ 7 $\boxed{=}$ $\boxed{=}$ $\boxed{=}$ Result:

Explain:

C. $\boxed{\text{A C}}$ 4 $\boxed{+}$ 3 $\boxed{+}$ 2 $\boxed{=}$ $\boxed{=}$ $\boxed{=}$ $\boxed{=}$ Result:

Explain:

D. $\boxed{\text{A C}}$ 45 $\boxed{-}$ 5 $\boxed{=}$ $\boxed{=}$ $\boxed{=}$ $\boxed{=}$ $\boxed{=}$ Result:

Explain:

E. $\boxed{\text{A C}}$ 60 $\boxed{-}$ 2 $\boxed{-}$ 3 $\boxed{=}$ $\boxed{=}$ $\boxed{=}$ Result:

Explain:

F. $\boxed{\text{A C}}$ 6 $\boxed{\times}$ $\boxed{=}$ $\boxed{=}$ $\boxed{=}$ $\boxed{=}$ Result:

Explain:

G. $\boxed{\text{A C}}$ 3 $\boxed{\times}$ 7 $\boxed{=}$ $\boxed{=}$ $\boxed{=}$ $\boxed{=}$ Result:

Explain:

H. $\boxed{\text{A C}}$ 2000 $\boxed{\div}$ 5 $\boxed{=}$ $\boxed{=}$ $\boxed{=}$ Result:

Explain:

I. $\boxed{\text{A C}}$ 35000 $\boxed{\div}$ 10 $\boxed{\div}$ 50 $\boxed{=}$ $\boxed{=}$ $\boxed{=}$ Result:

Explain:

NOTE: If your calculator does not operate properly in Part I, refer to the instructions for your specific model. It may have a $\boxed{\textbf{K}}$ key.

II. The $\boxed{y^X}$ Key

Complete this section if your calculator has a key marked $\boxed{y^X}$.

A.   1 $\boxed{y^X}$ 2 ⊟

B.   2 $\boxed{y^X}$ 1 ⊟

C.   2 $\boxed{y^X}$ 3 ⊟

D.   3 $\boxed{y^X}$ 2 ⊟

E.   3 $\boxed{y^X}$ 4 ⊟

F.   4 $\boxed{y^X}$ 3 ⊟

G.   4 $\boxed{y^X}$ 5 ⊟

H.   5 $\boxed{y^X}$ 4 ⊟

I.   5 $\boxed{y^X}$ 6 ⊟

J.   6 $\boxed{y^X}$ 7 ⊟

Summary: Compare $\mathbf{m}^n$ and $\mathbf{n}^m$, given that m < n. Which is larger?

ACTIVITY 2.2B - CHANGING BASES WITH A CALCULATOR

Materials required: Calculator

I. Base five

    A. In base five we are grouping by _ _ _ _ _ _ _ _ _ _ .

    B. Each place value is a power of _ _ _ _ _ _ _ _ _ _ .

    C. Whenever we have five pieces of one kind we must _ _ _ _ _ _ _ _ _ _ .

    D. The only allowable digits in base five are _ _ _ _ _ _ _ _ _ _ .

    E. Label the following place values in base five.

                                  _ _ _ _   _ _ _ _   _ _ _ _   _ _ _ _   _ _ _ _

                Power of 5   _ _ _ _   _ _ _ _   _ _ _ _   _ _ _ _   _ _ _ _

       Base 10 Equivalent   _ _ _ _   _ _ _ _   _ _ _ _   _ _ _ _   _ _ _ _

Example: Write 348 in base 5 notation.

    The first place value which is less than or equal to 348 is 125.

    To determine the number of 125's there are in 348 we use the following keystroke sequence.

        $\boxed{\text{A C}}$ 348 $\boxed{-}$ 125 $\boxed{=}$ $\boxed{=}$

    The number in the display is 98. This is less than 125 so now look for groups of 25.

        98   $\boxed{-}$ 25 $\boxed{=}$ $\boxed{=}$ $\boxed{=}$

    The number in the display is less than 25 so we now look for groups of 5.

        23   $\boxed{-}$ 5 $\boxed{=}$ $\boxed{=}$ $\boxed{=}$ $\boxed{=}$

    The number in the display is less than 5 so we know that it represents the number of units.

    Therefore, $348 = 2343_{\text{five}}$.

    Write each of the following in base five using your calculator.

    1. 59   2. 247   3. 2043

II. Change to base indicated using your calculator. Write the keystroke sequence you used.

    1. 47 to base 3

    2. 509 to base 4

    3. 1263 to base 6

    4. 2008 to base 7

    5. 1328 to base 9

ACTIVITY 2.2C-USING A CALCULATOR TO CHANGE BASES

Materials required: Calculator

$\boxed{M+}$ Adds the value in the display to that in the memory.

$\boxed{M-}$ Subtracts the value in the display from that in the memory.

$\boxed{MR}$ Recalls the value in the memory to the display.

Example: Write the base ten equivalent of 5123_{seven} .

METHOD I: Computation

$$\underline{5}\ \underline{1}\ \underline{2}\ \underline{3}_{\ seven} = 5(343) + 1(49) + 2(7) + 3(1) = 1781$$

$7^3\ 7^2\ 7^1\ 7^0$

METHOD II: Calculator

$\boxed{A\ C}$ 5 $\boxed{\times}$ 343 $\boxed{=}$ $\boxed{M+}$

1 $\boxed{\times}$ 49 $\boxed{=}$ $\boxed{M+}$

2 $\boxed{\times}$ 7 $\boxed{=}$ $\boxed{M+}$

3 $\boxed{\times}$ 1 $\boxed{=}$ $\boxed{M+}$ $\boxed{MR}$

1781 will be in the display.

Use your calculator to change the following numbers to base ten. Write the keystroke sequence you used.

1. 51_{seven}

2. 32_{four}

42

3. 543_{six}

4. 1013_{five}

5. 87_{nine}

6. 402_{five}

7. 2013_{four}

8. 543_{six}

9. 2137_{eight}

10. 2211_{three}

2.2 CHECK YOUR UNDERSTANDING

1. Give the result of the following keystroke sequence and explain the operations involved.

   $\boxed{\text{A C}}$ 3 $\boxed{\times}$ $\boxed{=}$ $\boxed{=}$ $\boxed{=}$ $\boxed{=}$ $\boxed{=}$

2. Give the result of the following keystroke sequence and explain the operations involved.

   $\boxed{\text{A C}}$ 2 $\boxed{+}$ 4 $\boxed{\times}$ $\boxed{=}$ $\boxed{=}$ $\boxed{=}$

3. Give the keystroke sequence you would use to change 286 to base 5 notation.

4. Give the keystroke sequence you would use to change 241_{seven} to base 10 notation.

3: ADDITION AND SUBTRACTION

Introduction: Through these activities we will explore the basic operations of "combination" (addition) and "separation" (subtraction). Models will be presented which demonstrate these operations.

I. Write word problems which could be solved using each of the equations below.

A. $5 + a = 26$

B. $12 - 8 = b$

C. $9 - c = 2$

D. $(8 + 3) - (5 + 1) = d$

II. Find one sale advertisement in a newspaper. Attach it in the space provided below. Write three word problems for the advertisement. One problem should require only addition in its solution, one problem should require only subtraction in its solution, and one problem should require BOTH addition and subtraction in its solution.

Advertisement:

Problem 1 (Addition only):

Problem 2 (Subtraction only):

Problem 3 (BOTH Addition and Subtraction):

ACTIVITY 3.1A - A MODEL FOR ADDITION

Materials required: Centimeter rods included with textbook

I. Which single rod is the same length as the train formed by the following rods?

    1.   C + E =

    2.   D + A =

    3.   B + G =

    4.   E + D =

    5.   I + A =

II. Complete the following chart. (NOTE: The A rod is 1 centimeter in length.)

ROD	LENGTH OF ROD IN CENTIMETERS
A	1
B	2
C	3
D	
E	
F	
G	
H	
I	
J	

III. In order to determine 6 + 3 we build a train using rod F, which is 6 cm long, and rod C, which is 3 cm long. We then look for a rod that is equal in length to that train. That rod is rod I. Since rod I is 9 cm long, 6 + 3 = 9.

Explain how one could determine the following sums using the centimeter rods.

    A.   4 + 3

    B.   8 + 2

IV. Complete the following chart. (Look for patterns which may save you time.)

+	A	B	C	D	E	F	G	H	I	J
A										
B										
C										
D										
E										
F										
G										
H										
I										
J										

V. Is this system commutative for addition? Does C + E = E + C ? Explain.

VI. Is this system associative for addition? Does B + (A + D) = (B + A) + D? Explain.

VII. Is the set of centimeter rods closed for addition? Explain.

50

ACTIVITY 3.1B - MODELS FOR SUBTRACTION

Materials *required*: *Centimeter rods included with textbook*

I. "Takeaway" Model.

    A. To determine I − E :

        1. Replace the I rod with centimeter rods.

        2. Remove the number of centimeter rods which equal the length of the E rod.

        3. Which rod is the same length as the centimeter rods that remain?

        This rod represents the difference.

    B. Perform each of the following subtractions using the Takeaway Model described above. Show your work as demonstrated in Part A.

        1.   J − H

        2.   F − D

        3.   G − B

        4.   I − C

        5.   I − C

C. Complete the following chart. (Look for patterns which may save time.)

−	A	B	C	D	E	F	G	H	I	J
A										
B										
C										
D										
E										
F										
G										
H										
I										
J										

D. Is this system commutative for subtraction? Does H − E = E − H? Explain.

E. Is this system associative for subtraction? Does G − (E − B) = (G − E) − B? Explain.

F. Is the set of centimeter rods closed for subtraction? Explain.

52

II. Missing Addend Model.

    A. To subtract I − E:

        Determine which rod can be added to the E rod to form a train which is equal in length to the I rod.

$$E + \_\_\_\_ = I$$

        This rod represents the difference.

    B. Perform each of the following subtractions using the missing addend model described above. Rewrite each of them as an addition problem. Show your work as demonstrated in Part A.

        1.   J − H

        2.   F − D

        3.   G − B

        4.   I − C

        5.   D − A

C. Demonstrate the difference, 8 − 3.

    1. Find the rod which is eight A rods in length. This is the _ _ _ _ _ _ rod.

    2. Find the rod which is three A rods in length. This is the _ _ _ _ _ _ rod.

    3. Compare the length of the two rods. Determine which rod may be added to the rod in number 2 to form a train equal in length to the rod in number 1. This is the _ _ _ _ _ _ rod.

    4. Determine the length of the rod in number 3 using A rods. The length is _ _ _ _ _ _ A rods. This is the difference.

D. Perform each of the following subtractions using the missing addend model described above.

    1. $\boxed{7-2}$ What must be added to the _ _ _ _ _ _ _ rod to form a train which is equal in length to the _ _ _ _ _ _ rod? This is the _ _ _ _ _ _ rod. It is _ _ _ _ _ _ A rods long. Therefore, $7 - 2 =$ _ _ _ _ _.

    2. $\boxed{9-4}$ What must be added to the _ _ _ _ _ _ rod to form a train which is equal in length to the _ _ _ _ _ _ rod? This is the _ _ _ _ _ _ rod. It is _ _ _ _ _ _ A rods long. Therefore, $9 - 4 =$ _ _ _ _ _.

    3. $\boxed{10-6}$ What must be added to the _ _ _ _ _ _ rod to form a train which is equal in length to the _ _ _ _ _ _ rod? This is the _ _ _ _ _ _ rod. It is _ _ _ _ _ _ A rods long. Therefore, $10 - 6 =$ _ _ _ _ _.

E. Explain how you would perform the following subtractions using the missing addend model.

    1. $7 - 5$

    2. $10 - 4$

3.1 CHECK YOUR UNDERSTANDING

1. Explain how one could use centimeter rods to perform the following operations.

    a. $7 + 3$

    b. $9 - 6$

2. Which of the following properties hold for our collection of centimeter rods? Explain.

    a. Associative property of subtraction

    b. Commutative property of addition

    c. Closure property of subtraction

ACTIVITY 3.2A - ADDITION USING BASE TEN PIECES

Materials required: Base ten pieces from the Appendix

EXAMPLE: Add 397 + 745 using base ten pieces.

   397 may be represented using 3 flats 9 longs and 7 units.

   745 may be represented using 7 flats 4 longs and 5 units.

   a. We have a total of 12 units. We must trade 10 units for one long.   Two units remain.

   b. We now have a total of 14 longs. We must trade 10 longs for one flat. Four longs remain.

   c. We now have 11 flats. We must trade 10 flats for one block.   One flat remains.

   d. The result is one block one flat four longs and two units which is written as 1142.

Use your base ten pieces to perform the following additions. Then complete the chart.

			B	F	L	U
1.	23 + 189	Before Trades:				
		After Trades:				
2.	796 + 64	Before Trades:				
		After Trades:				
3.	85 + 75	Before Trades:				
		After Trades:				
4.	683 + 948	Before Trades:				
		After Trades:				
5.	586 + 675	Before Trades:				
		After Trades:				

ACTIVITY 3.2B - SUBTRACTION USING BASE TEN PIECES

Materials Required: *Base ten pieces from the Appendix*

EXAMPLE: Subtract 835 − 649 using base ten pieces.

835 may be represented using 8 flats 3 longs and 5 units.

a. We must first remove 9 units. We must trade 1 long for 10 units. We now have 15 units. We now remove the 9 units. <u>Six</u> <u>units</u> remain.

b. We must remove 4 longs. We must trade 1 flat for 10 longs. We now have 12 longs. We now remove the 4 longs. <u>Eight</u> <u>longs</u> remain.

c. We must remove 6 flats. <u>One</u> <u>flat</u> remains.

d. The result is 1 flat 8 longs and 6 units. This may be written as 186.

Use your base ten pieces to perform the following subtractions.

		B	F	L	U
1. 48 − 29	Begin With:				
	Difference:				
2. 103 − 56	Begin With:				
	Difference:				
3. 224 − 187	Begin With:				
	Difference:				
4. 1001 − 897	Begin With:				
	Difference:				

ACTIVITY 3.2C-ADDITION IN BASE THREE

Materials required: _Base three pieces from the Appendix_

Example: Add 212_{three} + 222_{three} .

212_{three} may be represented using 2 flats 1 long 2 units.

222_{three} may be represented using 2 flats 2 longs 2 units.

a.   We have a total of 4 units. Three units must be traded for 1 long. One unit remains.

b.   We have a total of 4 longs. Three longs must be traded for 1 flat. One long remains.

c.   We have a total of 5 flats. Three flats must be traded for 1 block. Two flats remain.

d.   The result is 1 block 2 flats 1 long 1 unit which may be written as 1211_{three} .

Use your base three pieces to determine the following sums. Complete the chart.

		B	F	L	U
1. 20_{three} + 12_{three}	Before Trades: After Trades:				
2. 102_{three} + 221_{three}	Before Trades: After Trades:				
3. 202_{three} + 122_{three}	Before Trades: After Trades:				
4. 1011_{three} + 112_{three}	Before Trades: After Trades:				
5. 111_{three} + 1022_{three}	Before Trades: After Trades:				
6. 1002_{three} + 101_{three}	Before Trades: After Trades:				

59

3.2D - SUBTRACTION IN BASE THREE

Materials required: _Base three pieces from the Appendix_

Example: Subtract $221_{three} - 22_{three}$.

221_{three} may be represented as 2 flats 2 longs and 1 unit.

a.   In order to remove 2 units we must trade one long for 3 units. We now have 4 units and can subtract 2 units. <u>Two units remain</u>.

b.   In order to remove 2 longs we must trade one flat for 3 longs. We now have 4 longs and can subtract two longs. <u>Two longs remain</u>.

c.   The result is 1 flat 2 longs 2 units which may be written as 122_{three} .

Use your base three pieces to determine the following differences. Complete the chart.

			B	F	L	U
1.	$21_{three} - 12_{three}$	Begin with:				
		Difference:				
2.	$101_{three} - 22_{three}$	Begin with:				
		Difference:				
3.	$200_{three} - 11_{three}$	Begin with:				
		Difference:				
4.	$211_{three} - 2_{three}$	Begin with:				
		Difference:				
5.	$1000_{three} - 121_{three}$	Begin with:				
		Difference:				
6.	$1102_{three} - 201_{three}$	Begin with:				
		Difference:				

3.2 CHECK YOUR UNDERSTANDING

I. Perform the following operations using base ten pieces. Describe your work.

A.   489 + 732

B.   1024 − 537

II. Determine the following operations using base three pieces. Describe your work.

A.   211 $_{\text{three}}$ + 101 $_{\text{three}}$

B.   1100 $_{\text{three}}$ − 212 $_{\text{three}}$

4: MULTIPLICATION AND DIVISION

Introduction: Concrete models may be used to develop an understanding of multiplication and division of whole numbers. This understanding will promote the awareness of the properties which hold for these operations. This will facilitate the solution of problems encountered in daily life.

For each of the following equations give one situation which it may represent. You need not solve the equations.

1. $3 \times 5 = a$

2. $36 \div 9 = b$

3. $(8 \times 5) \div 2 = c$

4. $(12 \div 6) \times 3 = d$

ACTIVITY 4.1A - THE SET MODEL FOR MULTIPLICATION

Materials required: Centimeter rods included with the textbook

Example: 2 × 4 may be represented using centimeter rods.

Let the A rod represent one unit.

1. Which rod represents four units?
2. Which rod represents two of the rods in the previous question?
3. Represent the length of the rod in number 2 using A rods.
4. Therefore, 2 × 4 =

I. Perform each of the following multiplications using your centimeter rods. Be certain to give the letters of the rods used and the final answer.

A.   3 × 2 =

B.   2 × 5 =

C.   3 × 3 =

II. Sometimes we must use J rods and another letter rod in our answer.

A.   3 × 6 =

B.   2 × 7 =

C.   6 × 2 =

D.   5 × 4 =

III. Properties

    A. Does the commutative property of multiplication hold for the centimeter rods?

        For example, does 2 × 3 = 3 × 2? Explain.

    B. Does the associative property of multiplication hold for the centimeter rods?

        For example, does 2 × (6 × 3) = (2 × 6) × 3? Explain.

    C. Does the distributive property of multiplication hold for the centimeter rods?

        For example, does 3 (B + C) = 3 B + 3 C ? Explain.

ACTIVITY 4.1B - ARRAY MODEL FOR MULTIPLICATION

Materials required: _Square tiles included in the Appendix_

Example: 2 × 4 may be represented using square tiles.

We use our square tiles to build an array with two rows of four tiles each.

We then count our tiles.

There are eight tiles in our array.

Therefore, 2 × 4 = 8

I. Determine the following products using your square tiles. Sketch or describe the tiles used.

A.   3 × 2 =

B.   2 × 5 =

C.   3 × 3 =

D.   3 × 6 =

E. $2 \times 7 =$

F. $6 \times 2 =$

G. $5 \times 4 =$

II. Properties

    A. Does the commutative property of multiplication hold for the set of square tiles? For example, can we show that $2 \times 3 = 3 \times 2$? Explain.

    B. Does the associative property of multiplication hold for the set of square tiles? For example, can we show that $2 \times (3 \times 4) = (2 \times 3) \times 4$? Explain.

    C. Does the distributive property of multiplication hold for the set of square tiles? For example, does a 3×9 array have the same number of tiles as a 3×4 array added to a 3×5 array? Explain.

ACTIVITY 4.1C - SET MODEL FOR DIVISION

Materials required: Square tiles from the Appendix

Example: 10 ÷ 5 may be represented using square tiles.

Begin with a pile of 10 square tiles.

Sort them into 5 equal piles.

Each pile has 2 tiles in it.

Therefore, 10 ÷ 5 = 2.

I. Determine the following quotients using the set model described in the example above.

Provide a sketch or description of the process you use.

A.   12 ÷ 2 =

B.   20 ÷ 4 =

C.   21 ÷ 3 =

D.   15 ÷ 5 =

II. Sometimes we have tiles left over. When there are not enough to distribute evenly, we count the tiles and consider the number our remainder.

  A. 20 ÷ 6 =

  B. 26 ÷ 4 =

  C. 10 ÷ 3 =

  D. 27 ÷ 5 =

  E. 23 ÷ 7 =

  F. 30 ÷ 4 =

ACTIVITY 4.1D - MISSING FACTOR MODEL FOR DIVISION

Materials required: Square tiles from the Appendix, centimeter rods included with the textbook

Example: $10 \div 2 = n$ may be rewritten as $2 \times n = 10$.

I. Using square tiles:

We must build an array with two rows using ten tiles. When we arrange
the ten tiles into two rows we will have 5 tiles in each row.
Since $2 \times 5 = 10$, we know that $10 \div 2 = 5$.

Determine the following quotients using your square tiles. Write the related multiplication
example and describe or sketch your work.

A. $12 \div 4 = n$

B. $15 \div 5 = n$

C. $21 \div 3 = n$

D. $18 \div 2 = n$

II. Using centimeter rods:

> Begin with the J rod because it is ten A rods in length.

> The B rod is two A rods in length.

> We compare the J rod and the B rod.

> We determine that five B rods equal the length of the J rod.

> Since $5 \times 2 = 10$, we know that $10 \div 2 = 5$.

Determine the following quotients using your centimeter rods. Write the related multiplication example and describe or sketch your work.

A. $6 \div 3 = n$

B. $8 \div 2 = n$

C. $12 \div 4 = n$

       (Hint: Represent 12 as the J rod plus the B rod.)

D. $18 \div 9 = n$

4.1 CHECK YOUR UNDERSTANDING

1. Explain how you could determine the product, 5 × 3 , using a set model.

2. Explain how you could determine the product, 5 × 3 , using an array model.

3. Explain how you could determine the quotient, 24 ÷ 4 , using a set model.

4. Explain how you could determine the quotient, 24 ÷ 4, using an array model.

ACTIVITY 4.2 - EXPLAINING OTHER ALGORITHMS

I. Multiplication

   A. Estimate the product 192 × 376.

   B. Determine the product 192 × 376 using each of the following algorithms.

      1. Lattice Multiplication

      2. Partial Products

   C. Are the algorithms in Part B related? Explain.

   D. In what way are these algorithms related to the standard algorithm?

II. Division

    A. Estimate the quotient $450 \div 18$.

    B. Determine the quotient $450 \div 18$ using each of the following algorithms.

        1. Egyptian Division Method

        2. Subtractive Method

    C. Are the two algorithms in Part B related? Explain.

    D. In what way are these algorithms related to the standard algorithm?

4.2 CHECK YOUR UNDERSTANDING

1. a. Multiply 234 × 157 using a nonstandard algorithm.

    b. Explain the relationship between the algorithm you used in part a and the standard multiplication algorithm.

2. a. Divide 322 by 14 using a nonstandard algorithm.

    b. Explain the relationship between the algorithm you used in part a and the standard division algorithm.

ACTIVITY 4.3A - FACTORS & DIVISIBILITY

Materials required: *Square tiles from the Appendix*

1. Given 12 square tiles determine all the possible rectangular arrangements.

    a. Give the dimensions of each arrangement.

    b. Each pair of dimensions in Part A represents a pair of factors of 12. List the factors of 12 below.

    c. We say that each of the factors in Part B divides 12, or 12 is divisible by each of the factors. If a divides b, then we write a | b. Use this notation with each of the factors.

2. Use the method outlined above to determine the factors of each of the following numbers. List the factors and then use the proper notation to show that these factors divide the number.

    a.   8

    b.   7

    c.   20

    d.   16

3. a. Which of the numbers in #2 had more than two possible rectangles? These are composite numbers.

   b. Name another composite number and demonstrate that it is composite using square tiles.

4. a. Did any of the numbers in #2 have <u>only</u> <u>two</u> possible rectangles? These are prime numbers.

   b. Name another prime number and demonstrate that it is prime using square tiles.

5. a. How could you use square tiles to demonstrate that 15 and 14 are relatively prime?

   b. How could you use square tiles to demonstrate that 8 and 12 are not relatively prime?

   c. Name another pair of numbers that are relatively prime and demonstrate this using square tiles.

ACTIVITY 4.3B - THE SIEVE OF ERATOSTHENES

1	2	3	4	5	6	7	8	9	10
11	12	13	14	15	16	17	18	19	20
21	22	23	24	25	26	27	28	29	30
31	32	33	34	35	36	37	38	39	40
41	42	43	44	45	46	47	48	49	50
51	52	53	54	55	56	57	58	59	60
61	62	63	64	65	66	67	68	69	70
71	72	73	74	75	76	77	78	79	80
81	82	83	84	85	86	87	88	89	90
91	92	93	94	95	96	97	98	99	100

1. Since 1 is neither prime nor composite, put a triangle around it.

2. Put a circle around 2. Two is prime. Cross out all of the multiples of 2. They are composite.

3. Put a circle around 3. Three is prime. Cross out all of the multiples of 3. They are composite.

4. Four is crossed out. Therefore, we circle 5. Five is prime. Cross out all of the multiples of 5. They are composite.

5. Six is crossed out. Therefore, we circle 7. Seven is prime. Cross out all of the multiples of 7. They are composite.

6. What is the next number that has not been crossed out? Are its multiples already crossed out?

7. Circle the numbers which remain. List them below. They are the prime numbers between one and one hundred.

4.3 CHECK YOUR UNDERSTANDING

1. Explain the way in which to determine all the factors of 24 using square tiles.

2. Is 21 a prime number? How would you demonstrate this?

3. Are 6 and 10 relatively prime? How could you demonstrate this using square tiles?

4. Explain how the Sieve of Eratosthenes can be used to determine all the prime numbers from 1 to 100 inclusive.

ACTIVITY4.4A - EQUIVALENCE RELATIONS

Materials required: Attribute blocks from the Appendix, centimeter rods included with textbook

Determine if the given relation and set define an equivalence relation.

If YES, then give any partitions into which the set is divided.

If NO, then list each property which is violated (reflexive, symmetric, transitive) and give an example of the violation.

1. Centimeter rods, "is two A rods shorter than"

2. Centimeter rods, "is longer than"

3. Centimeter rods, "is half as long as"

4. Attribute blocks, "is the same color as"

5. Attribute blocks, "is smaller than"

6. Attribute blocks, "has the same number of sides as"

ACTIVITY 4.4B - FUNCTIONS

Materials required: Function machine visuals from the Appendix.

I. A. Determine the day on which you were born by using the following method.

    1. Write the tens and ones digits of the year you were born.   _ _ _ _ _ _

    2. Divide this number by 4 and drop the remainder. Write the   _ _ _ _ _ _
quotient without the remainder.

    3. Find the code number for your birth month in the chart below.   _ _ _ _ _ _
Write it here.

January	1	July	0
February	4	August	3
March	4	September	6
April	0	October	1
May	2	November	4
June	5	December	6

    4. Write the date of the month on which you were born.   _ _ _ _ _ _

    5. Add the numbers in steps 1 through 4.   _ _ _ _ _ _

    6. Divide the sum in step 5 by 7. The remainder tells the day of the week on which you were
born.   1-Sunday, 2-Monday, 3-Tuesday, 4-Wednesday, 5-Thursday, 6-Friday, 7-Saturday.
(NOTE: This method is only for dates in the 20th century.)

  B. On what day were you born?

  C. Ask seven of your classmates for their responses to B. Record them below by listing them in
the form (Name, Day).

  D. Is this a function? Why?
If so, what is the domain?
If so, what is the range?

  E. Record the results in Part C as (Day, Name). Is this a function? Explain.
If so, what is the domain?
If so, what is the range?

II. Composition of Functions

    A. Let **f** be the function which assigns to a given number its positive square root. Assume the domain of **f** is the set { 1, 4, 9, 16, 25, 36, 49, 64, 81, ... }. What is the range of **f**?

       Using your function machine visuals explain the meaning of **f** ∘ **f** . Is this defined? Explain.

    B. Let **g** be the function which assigns to a given number the square of the number. Assume the domain of **g** is the set { 2, 4, 6, 8, 10, ... }. What is the range of **g** ?

       Using your function machine visuals explain the meaning of **g** ∘ **g** . Is this defined? Explain.

    C. Use your function machine visuals to explain the meaning of each of the following using the functions described in parts A and B. Is the composition defined? Explain.

       1. **f** ∘ **g**

       2. **g** ∘ **f**

ACTIVITY 4.4C - CLOCK ARITHMETIC

Materials required: One brass brad, clock 12 card from the Appendix.

I. Addition

   A. Complete the following table using your clock 12 card with a brass brad attached at the center to form a pointer.

$\oplus$	1	2	3	4	5	6	7	8	9	10	11	12
1												
2												
3												
4												
5												
6												
7												
8												
9												
10												
11												
12												

   B. Which of the following properties hold for $\oplus$ in Clock 12? Explain.

      1. Commutative

      2. Associative

      3. Closure

   C. Is there an identity element for $\oplus$? Explain.

   D. Does each element have an inverse for $\oplus$? Explain.

II. Subtraction

    A. Using the table for $\oplus$ explain how the difference $4 \ominus 8$ could be determined.

    B. Is there closure for $\ominus$? Explain.

III. Multiplication

    A. Complete the following table using your clock 12 card with a brass brad attached at the center to form a pointer.

$\otimes$	1	2	3	4	5	6	7	8	9	10	11	12
1												
2												
3												
4												
5												
6												
7												
8												
9												
10												
11												
12												

    B. Which of the following properties hold for $\otimes$ in clock 12? Explain.

        1. Commutative

2. Associative

3. Closure

C. Is there an identity element for $\otimes$? Explain.

D. Does each element have an inverse for $\otimes$? Explain.

IV. Division

A. Using the table for $\otimes$ explain how the quotient $3 \oslash 5$ is determined.

(Hint: Determine a value for n which makes the statement, $5n = 3$, true.)

B. Is there closure for $\oslash$? Explain.

ACTIVITY 4.4D - MODULO THREE

Materials required: One brass brad, modulo three card from the Appendix

Robert, Shanikwa, and Tracy take turns babysitting little Zunilda.
They always rotate their duties. Robert babysits one day, Shanikwa
the next, and Tracy the next. We can represent this situation using
modulo three.

Place a brass brad through the center of the circle on the Modulo Three Card to
form a pointer.

I. Addition

   A. Complete the modulo three addition table by moving the brass brad the given number of
       of units.

+	0	1	2
0			
1			
2			

   B. If we know that Robert babysat today, predict who will babysit 2 days from now.

       (Let 0, 1, and 2 represent Robert, Shanikwa, and Tracy respectively.)

   C. If we know that Tracy babysat today, predict who will babysit 3 days from now.

   D. If we know that Shanikwa babysat today, could we use this model to predict which
       person will babysit ten days from now? Explain.

   E. Explain the way in which the answer in Part D relates to the sum of 11.

II. Subtraction

   A. Complete the subtraction table below by using your modulo three card to determine the number
       of units you must turn counterclockwise from the number at left to arrive at the one on the top.

−	0	1	2
0			
1			
2			

91

B. If we know Shanukwa babysat today, determine who babysat 2 days ago.

C. Write a problem for our situation which could be represented as $0 - 3 = n$.

III. Multiplication

A. Complete the multiplication table below by moving the brass brad on your modulo three card. For example, 2×2 means that beginning at zero you move the brad two turns of two units each.

X	0	1	2
0			
1			
2			

B. Examine your table. Which of the following properties hold for modulo three? Explain.

1. Commutative

2. Associative

3. Closure

IV. Division

Is division defined for modulo three? For example, how might we determine the quotient $1 \div 2$?

4.4 CHECK YOUR UNDERSTANDING

1. Determine if the given relation and set define an equivalence relation.

   If YES, then give any partitions into which the set is divided.

   If NO, then list each property which is violated (reflexive, symmetric, transitive) and

   give an example of the violation.

   Relation: "is twice as long as"

   Set: centimeter rods

2. Function **f** assigns each even whole number to the product of that number and four.

   Function **g** assigns each whole number greater than seven to the sum of that number and five.

   a. Give the range of each function.

   range of **f**:

   range of **g**:

   b. Is **g** ∘ **f** defined? Explain.

3. Determine each of the following on the Clock 12 set {1, 2, 3, 4, 5, 6, 7, 8, 9, 10, 11, 12}.

   a. $9 \oplus 5$ b. $3 \ominus 8$ c. $4 \otimes 6$ d. $5 \oslash 3$

4. Determine the value of B in each of the following.

   a. $1 + 5 = B \pmod 3$ b. $2 - 4 = B \pmod 3$

5: INTEGERS

Introduction: An understanding of the operations on the whole numbers and their opposites is an important part of our understanding of the world. This understanding will increase our ability to represent everyday situations and to solve problems accurately and confidently. The activities in this section are meant to provide a foundation for this understanding.

1. Find one application of negative integers in a newspaper, magazine, textbook, or other source.

2. Attach a copy of this application in the space provided below.

3. What is one way in which this situation could have been described without using negative integers?

4. How do negative integers improve our ability to describe the situation above?

ACTIVITY 5.1A - INTEGER ADDITION

Materials required: *Chips and box diagram from the Appendix*

EXAMPLE: The set model for integer addition uses chips marked with $-$ and $+$. In the following diagram, the addition, $(+5) + (-5)$, is illustrated.

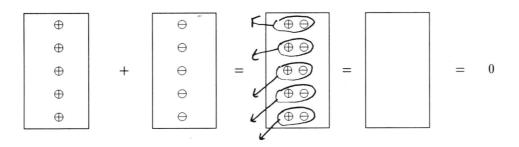

1. Add each of the following using chips.

   a. $(-4) + (+4)$ b. $(+4) + (-4)$

     Result: _ _ _ _ _   Result: _ _ _ _ _

2. a. Does $(-4) + (+4) = (+4) + (-4)$?

   b. Will $(-x) + (+x) = (+x) + (-x)$ always be true? Explain.

   c. What property have we demonstrated here?

3. Determine the following sums using the chips and the box diagram provided. Then illustrate your work in the space below each exercise. An example is done for you.

EXAMPLE: $(-7) + (+8)$

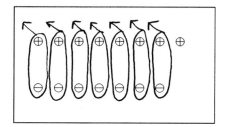

Result: _ _ _ _ _

a. $(+3) + (-5)$

Result: _ _ _ _ _

b. $(-6) + (-3)$

Result: _ _ _ _ _

c. $(-2) + (+4)$

Result: _ _ _ _ _

d. $(+8) + (-1)$

Result : _ _ _ _ _

98

ACTIVITY 5.1B - INTEGER SUBTRACTION

Materials required: Chips and box diagram from the Appendix

I. SET MODEL

EXAMPLE: The set model may be used to find the difference, $(-4) - (-1)$. Begin with four negative chips and remove one negative chip. Three negative chips remain so the result is -3. This operation is illustrated below.

A. Interpret $(-6) - (+2)$ using the set model.

1. What must be done in order to perform this subtraction?

2. How can this be accomplished without changing the value of -6?

3. What is the result?

4. Illustrate the set model below.

B. Explain how the set model could be used to determine each difference. Sketch the operation.

1. $(-7) - (+4)$

Result: _ _ _ _ _

2. $(+8) - (+3)$

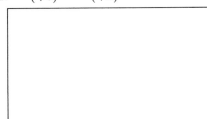

Result: _ _ _ _ _

3. (+2) − (+4)

Result: _ _ _ _ _

4. (+9) − (−1)

Result: _ _ _ _ _

5. (− 3) − (− 6)

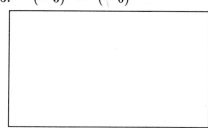

Result: _ _ _ _ _

6. (− 4) − (+ 5)

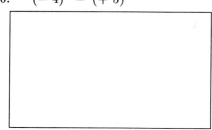

Result: _ _ _ _ _

II. ADD-THE-OPPOSITE

EXAMPLE: To find the difference, (−4) − (−1), we could add one positive chip instead of removing one negative chip. As seen below, the positive chip cancels one of the negative chips and we have three negative chips remaining. The result is still −3.

 → 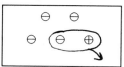 →

Determine each of the following differences using this approach. Sketch your work.

1. (−6) − (+2)

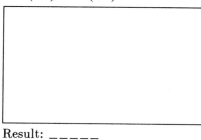

Result: _ _ _ _ _

2. (−7) − (+4)

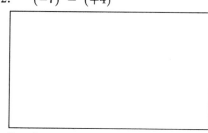

Result: _ _ _ _ _

100

3. $(+8) - (-3)$

Result: _ _ _ _ _

4. $(+2) - (+4)$

Result: _ _ _ _ _

5. $(+9) - (-1)$

Result: _ _ _ _ _

6. $(-3) - (-6)$

Result: _ _ _ _ _

III. MISSING ADDEND MODEL

EXAMPLE: The equation, $(-4) - (-1) = n$, can be rewritten as $(-1) + n = (-4)$.

We may ask, "What must be added to one negative chip in order to have four negative chips?"

The answer is "three negative chips." The difference is -3.

Use this method to determine each of the following differences. Give the question you would ask and the answer.

1. $(-6) - (+2) = n$

2. $(-7) - (+4) = n$

3. $(+8) - (+3) = n$

4. $(+2) - (+4) = n$

5. $(+9) - (-1) = n$

6. $(-3) - (-6) = n$

101

5.1 CHECK YOUR UNDERSTANDING

1. Explain how you could use chips to represent -4 in three different ways.

   a.

   b.

   c.

2. Explain how you can use chips to determine the sum, $(+4) + (-7)$.

3. Explain how you can use the Chip Model to determine the difference, $(-8) - (+3)$.

4. Explain how you can use Add-the-Opposite to determine the difference, $(+4) - (-2)$.

5. Explain how you can use the Missing Addend Approach to determine the difference, $(-6) - (-4)$.

ACTIVITY 5.2A - INTEGER MULTIPLICATION

Materials required: *Chips and box diagram from the Appendix*

EXAMPLE: The product $(+4) \cdot (+3)$ can be determined using a chip model. It may be thought of as $(+3) + (+3) + (+3) + (+3)$. Therefore, we begin with the empty set and add four groups of three positive chips. Since the result is 12 positive chips the product is $+12$. The diagram below shows this operation.

 $\rightarrow$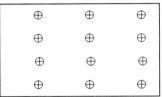

1. Given: $(+4) \cdot (-3)$

   a. Rewrite the product as a sum.

   b. Interpret the product using chips.

   c. Record your work in the diagram below.

   Result: _ _ _ _ _

2. Given: $(-4) \cdot (+3)$

   a. In its present form does this represent addition?

   b. What operation does the -4 represent?

   c. Can we perform this operation if we begin with the empty set?

   d. Rename the empty set so the operation may be performed. What chips must you use?
      (Remember: You must begin with a set with a value of zero.)

   e. Now perform the operation. What is the result?

   f. Sketch your work below. Explain.

105

3. Determine the product of $(-4) \cdot (-3)$ using chips. Sketch your work and explain.

4. Determine each product using chips. Sketch your work below each problem.

a. $(+3) \cdot (-5)$

Result: _ _ _ _ _

b. $(-6) \cdot (-1)$

Result: _ _ _ _ _

c. $(-8) \cdot (-2)$

Result: _ _ _ _ _

d. $(-2) \cdot (+4)$

Result: _ _ _ _ _

e. $(+7) \cdot (+2)$

Result: _ _ _ _ _

f. $(-2) \cdot (+5)$

Result: _ _ _ _ _

ACTIVITY 5.2B - INTEGER DIVISION

Materials required: Chips and box diagram from the Appendix

EXAMPLE: Consider the example, $(+15) \div (+5) = n$. By the definition of division, $(+5) \cdot n = (+15)$. The missing factor, n, may be determined using chips. Begin with the empty set. Add five equivalent groups of chips for a resulting value of $+15$. How many chips are in each equivalent group? This procedure is illustrated in the diagram below.

1. Given: $(+8) \div (-2) = n$

   a. Rewrite the equation using multiplication.

   b. Consider the meaning of -2 in the equation which you have written in part a. Will you add equivalent groups or subtract equivalent groups?

   c. With how many equivalent groups will you operate?

   d. What must be the result of the operation?

   e. Begin with an empty box diagram. This represents zero. How must you change this set in order to perform the operation implied by your statement in part a?

   f. Perform the operation. Represent your work in the box below.

   g. How many chips were in each of the equivalent sets?

   h. What is the value of each equivalent set?

   i. What is the quotient of $(+8) \div (-2)$?

107

2. Determine each of the following quotients using your chips and box diagram. Then represent the procedure below each problem.

a. $(+6) \div (+3)$

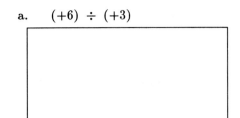

    Result: _ _ _ _ _

b. $(+4) \div (-2)$

    Result: _ _ _ _ _

c. $(-12) \div (-3)$

    Result: _ _ _ _ _

d. $(-10) \div (+5)$

    Result: _ _ _ _ _

e. $(-8) \div (-4)$

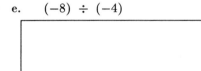

    Result: _ _ _ _ _

f. $(+6) \div (-2)$

    Result: _ _ _ _ _

3. Is integer division associative? Explain.

4. Is integer division commutative? Explain.

5.2 CHECK YOUR UNDERSTANDING

1. Given: $(-3) \cdot (-2) = m$

   a. Determine the product using the chip model.

   b. Represent your work below.

   c. Explain the procedure that you used in the diagram above.

2. Interpret $(-12) \cdot 0 = 0 \cdot (-12)$ using a chip model.

3. Could the product $(-7) \cdot (+2)$ be determined using repeated addition? What property would allow you to rewrite the problem in this way?

4. Justify $(+2) [(+3) + (-1)] = (+2)(+3) + (+2)(-1)$ using the chip model.

5. How can you determine the product $(-200) \cdot (+42)$ without using chips?

6. Given: $(+10) \div (-2) = n$

   a. Determine the quotient using the chip model.

   b. Represent your work below.

   c. Explain the procedure that you used in the diagram above.

7. Given: $(-9) \div (+4)$

   a. Can this quotient be determined using the chip model? Explain.

   b. How does this affect closure for integer division?

8. How can we determine the quotient, $(-100) \div (-25)$, without using a chip model? Explain your answer.

ACTIVITY 5.3 - INTRODUCTION TO ALGEBRA

Materials required: Chips and box diagram from the Appendix

EXAMPLE: In the equation, x + 2 = 5, the x represents the number of chips in set X which must be added to two positive chips in order to have a set of five positive chips. To determine the number of chips represented by set X we remove the two positive chips from the set X + A. We must also remove two positive chips from set B to maintain equivalence. There are three positive chips remaining in set B. Therefore, x represents three positive chips. This operation is illustrated in the diagram below.

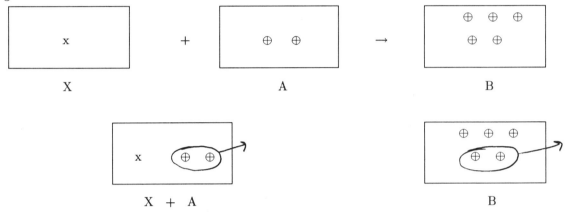

x + 2 = 5 becomes (x + 2) − 2 = 5 − 2

A. EQUATIONS

Use chips and the properties of equality to determine the solutions to the following equations.

a. Interpret the meaning of x in each equation.

b. Explain the process using chips.

c. Represent the process by rewriting the equation.

1. x + 3 = 4

2. x − 2 = −4

[Hint: Rewrite the equation as x + (−2) = −4]

3. $2x = -6$ 4. $x \div 4 = -2$

5. $2x + 4 = 12$

B. INEQUALITIES

Explain how the chip model could be applied to each of the following. Show your work.

1. $3x + 1 > 7$ 2. $x + 4 > 9$

3. $2x - 3 < 5$

5.3 CHECK YOUR UNDERSTANDING

1. Determine the solution of x − 5 = 6 using a chip model. Explain.

2. Determine the solution of x ÷ 2 = −3 using a chip model. Explain.

3. Find the solution of 2x + 3 = 7 using a chip model. Explain.

4. Solve 4x − 1 > 11 using a chip model. Explain.

6: RATIONAL NUMBERS

Introduction: The ability to use rational numbers is important in a world where we must often compare parts to wholes and parts to parts. Measurements and proportions require an understanding of these numbers. The activities in this section will help develop this understanding.

I. Find ten ways in which fractions are used in magazines, newspapers, and advertisements. List the contexts in which they are used.

   1.

   2.

   3.

   4.

   5.

   6.

   7.

   8.

   9.

   10.

II. A. Select a favorite recipe. How many people does it serve?

   B. Suppose you were expecting $\frac{2}{3}$ as many people for dinner. How many people would this be?

   C. Suppose you were expecting twenty people for dinner. In what way would you alter the recipe in order for everyone to have enough to eat?

ACTIVITY 6.1A - AN AREA MODEL FOR FRACTIONS

Materials required: *Fraction circles from the Appendix, tangrams from the Appendix*

I. For each circle:

    A. Count the number of pieces into which the circle is divided.

    B. What fraction of the circle does each piece represent?

    C. Label each piece.

    D. Cut out the pieces.

II. Write the following fractions in increasing order. Use your fraction pieces to compare them.

$$\frac{1}{6} \quad \frac{1}{2} \quad \frac{1}{8} \quad \frac{1}{3} \quad \frac{1}{4}$$

    If the numerator remains constant and the denominator is increased, the resulting fraction

    is _____ than the original fraction.

III.A. Represent $\frac{3}{4}$ using your pieces. How many eighths will completely cover them?

$$\text{So, } \frac{3}{4} = \frac{}{8} \ .$$

    B. How many twelfths are equal to three fourths? How could you show this using your fraction pieces?

IV.A. Represent $\frac{2}{3}$ using your fraction pieces. How many sixths will completely cover them?

$$\text{So, } \frac{2}{3} = \frac{}{6} \ .$$

    B. How many twelfths are equal to two thirds? How could you show this using your fraction pieces?

V. In the set of tangrams, let the large triangle represent one whole. What fraction does each of the following represent?

   A. the small triangle

   B. the medium triangle

   C. the square

   D. the rhombus

VI. In the set of tangrams, let the square represent one whole. What fraction does each of the following represent?

   A. the small triangle

   B. the medium triangle

   C. the large triangle

   D. the rhombus

VII. In the set of tangrams, let the medium triangle represent one whole. What fraction does each of the following represent?

   A. the small triangle

   B. the large triangle

   C. the rhombus

   D. the square

ACTIVITY 6.1B-COMPARING FRACTIONS WITH CENTIMETER RODS

__Materials required__: *Centimeter rods included with the textbook*

I. Begin with a J rod and a B rod. Let these rods together define one whole.

A. Fractions

    1. a. How many A rods form a "train" equal in length to the J and B rods?

        b. Each A rod is what part of a whole?

    2. a. How many B rods form a "train" equal in length to the J and B rods?

        b. Each B rod is what part of a whole?

    3. a. How many C rods form a "train" equal in length to the J and B rods?

        b. Each C rod is what part of a whole?

    4. a. How many D rods form a "train" equal in length to the J and B rods?

        b. Each D rod is what part of a whole?

    5. a. How many F rods form a "train" equal in length to the J and B rods?

        b. Each F rod is what part of a whole?

B. Order the fractions you wrote in numbers 1 through 5 from least to greatest. Do you notice a pattern? Summarize.

C. Compare $\frac{5}{6}$ and $\frac{3}{4}$ using the model above.

D. Compare $\frac{2}{3}$ and $\frac{5}{12}$ using the model above.

II. Begin with a "train" of one J rod and one H rod. Let this represent one whole.

    A. Writing fractions. What fraction does each of the following rods represent?

        1. A rod

        2. B rod

        3. C rod

        4. F rod

        5. I rod

    B. Order the fractions which you wrote in Part A from least to greatest.

    C. Comparison.

        Use your rods to determine the greater of the two fractions and circle it.

        1. $\frac{2}{9}$; $\frac{1}{3}$

        2. $\frac{3}{18}$; $\frac{1}{6}$

        3. $\frac{1}{2}$; $\frac{3}{9}$

III. Combine your set of rods with those of a classmate. Begin with two J rods and one D rod. Let this represent one whole.

    A. What fraction does each of the following represent?

        1. A rod

        2. B rod

        3. C rod

        4. D rod

        5. F rod

        6. H rod

    B. Compare the following fractions. Circle the greater of the two fractions.

        1. $\frac{2}{3}$; $\frac{5}{8}$

        2. $\frac{3}{8}$; $\frac{5}{12}$

        3. $\frac{5}{6}$; $\frac{7}{8}$

ACTIVITY 6.1C - THE DENSITY PROPERTY

I. A. How many rational numbers are between $\frac{1}{4}$ and $\frac{1}{5}$?

    B. Rewrite $\frac{1}{4}$ and $\frac{1}{5}$ with a common denominator.

    C. Can you name a rational number between the two fractions in Part B?

    D. Rewrite the fractions in Part B with denominators of 1000.

    E. Now can you name a rational number between the fractions in Part D?

    F. Could you have rewritten in Part D using a larger denominator?

    G. Do you agree with your original answer to Part A?

    H. The Density Property states that between any pair of fractions there is another fraction.

II. Find a rational number between each of the following pairs of rational numbers using the method in Part I.

    1. $\frac{1}{14}$; $\frac{3}{10}$

    2. $\frac{2}{7}$; $\frac{1}{6}$

    3. $\frac{1}{8}$; $\frac{1}{9}$

    4. $\frac{1}{11}$; $\frac{1}{12}$

    5. $\frac{1}{30}$; $\frac{1}{31}$

III.A. Determine the sum of the numerators and the sum of the denominators of $\frac{1}{4}$ and $\frac{1}{5}$.

B. Compare the fractions given in Part A to the result of the addition in Part A. Which fraction is between the other two?

C. Now determine the sum of the numerators and the sum of the denominators of $\frac{1}{4}$ and $\frac{2}{9}$.

D. Compare the two fractions given in Part C with the result of the addition. Which fraction is between the other two?

IV. Use the method in Part III to determine a rational number between each pair of rational numbers.

1. $\frac{1}{14}$; $\frac{3}{10}$

2. $\frac{2}{7}$; $\frac{1}{6}$

3. $\frac{1}{8}$; $\frac{1}{9}$

4. $\frac{1}{11}$; $\frac{1}{12}$

5. $\frac{1}{30}$; $\frac{1}{31}$

V. Are the results for Parts II and IV identical? Explain.

ACTIVITY 6.1D - LCM WITH CENTIMETER RODS

Materials required: Centimeter rods included with the textbook.

EXAMPLE: Determine the least common multiple of 4 and 6. Form a "train" of D rods (four units in length) and a "train" of F rods(six units in length). These "trains" represent multiples of the numbers. Continue adding rods until there is a common break. Their common break occurs after two F rods(twelve units) and three D rods(twelve units). Therefore, the least common multiple is 12.

Use the method described above to determine the LCM(least common multiple) of the following numbers. Describe or sketch your work. You may need to duplicate or trace extra rods.

1.   6; 8

2.   4; 10

3.   3; 5

4.   8; 3

5.   3; 7

6.   4; 8; 3

NAME _____ DATE _____

6.1 CHECK YOUR UNDERSTANDING

1. Explain the meaning of the fraction, $\frac{3}{8}$, using an area model.

2. Using centimeter rods explain which is greater, $\frac{3}{4}$ or $\frac{2}{3}$?

3. What is the Density Property?

4. List three fractions between $\frac{1}{7}$ and $\frac{1}{6}$.

5. Determine the LCM of 6 and 9 using centimeter rods.

ACTIVITY 6.2A-ADDITION OF RATIONAL NUMBERS

Materials __required__: Centimeter rods included with textbook

I. Represent one whole as a "train" consisting of a J rod and a B rod.

    A. Make all possible one letter trains which are equal in length to the J and B rod train.

    B. Represent each of the letters used in Part A as a fraction of the whole.

        Letter   Fraction of Whole

    C. In order to determine the sum, $\frac{1}{3} + \frac{5}{6}$, we must represent $\frac{1}{3}$ as sixths. Explain how this may be done and the way in which it leads to the sum.

    D. In order to determine the sum, $\frac{3}{4} + \frac{1}{12}$, we must represent $\frac{3}{4}$ as twelfths. Explain how this may be done and the way in which it leads to the sum.

II. Suppose that we wanted to show $\frac{3}{5} + \frac{1}{10}$. What rod, or rods, could we use as one whole? Explain how this addition could be performed.

III.In order to perform the addition, $\frac{1}{20} + \frac{4}{5}$, what could we use as one whole? Explain how this addition may be performed.

127

ACTIVITY 6.2B-SUBTRACTION OF RATIONAL NUMBERS

Materials _required_: _Centimeter rods included with textbook_

I. Let a "train" consisting of one J rod and one B rod represent one whole.

    A. Example: Determine the difference, $\frac{2}{3} - \frac{1}{6}$, using centimeter rods.

        1. Find the rod that is $\frac{1}{3}$ the length of the J and B "train." It is the D rod. Therefore, two D rods may be used to represent $\frac{2}{3}$.

        2. Find the rod that is $\frac{1}{6}$ the length of the J and B "train." It is the B rod.

        3. Compare the two D rods to the B rod. What rod must be added to the B rod to result in a "train" which is equal in length to the two D rods? It is the F rod.

        4. How many F rods are needed to build a "train" as long as the J and B "train?" Two F rods are equal in length to the J and B "train." Each F rod must be $\frac{1}{2}$.

        5. Therefore, the difference is $\frac{1}{2}$.

    B. Explain how one could determine the following differences using the centimeter rods.

        1. $\frac{5}{6} - \frac{7}{12}$

        2. $\frac{3}{4} - \frac{1}{3}$

        3. $\frac{7}{12} - \frac{1}{2}$

II. Suppose that we want to demonstrate $2 - \frac{1}{7}$. What rod, or rods, could we use to represent one whole? Explain how we could find the difference.

6.2 CHECK YOUR UNDERSTANDING

I. Let the J and B rods form a "train" which represents one whole.

Explain how you could determine each of the following using centimeter rods.

1. $\dfrac{5}{12} + \dfrac{1}{2}$

2. $\dfrac{1}{4} - \dfrac{1}{6}$

3. $2 - \dfrac{5}{6}$

II. Suppose we wanted to show $\dfrac{5}{8} - \dfrac{1}{4}$. What rod, or rods, could we use as one whole? Explain how we could determine the difference.

ACTIVITY 6.3A-WHOLE NUMBER MULTIPLIERS

Materials *required*: *Centimeter rods included with textbook*

Products involving whole number multipliers may be determined using centimeter rods. For example, $5 \times \frac{1}{3}$ may be interpreted as five rods each of which is one third of a whole. Let the "train" which consists of the J and B rods represent one whole. The D rod represents one third since three D rods are equal in length to the J and B "train." Five D rods will, therefore, be equal to one whole with two D rods left over. The two D rods represent $\frac{2}{3}$ of a whole. The product is $1\frac{2}{3}$.

I. Let the J and B rods represent one whole. Determine each of the following products using the centimeter rods. Explain your work.

1. $\quad 3 \times \frac{1}{2}$

2. $\quad 6 \times \frac{1}{4}$

3. $\quad 16 \times \frac{1}{12}$

II. Form a group of four people with three of your classmates. Explain how one could determine the product, $4 \times 2\frac{1}{3}$, using the centimeter rods.

ACTIVITY 6.3B-FRACTION MULTIPLIERS

__Materials__ __required__: Paper rectangles, pencil and pen (or two different colored pens or pencils)

When the multiplier is a fraction an array model may be used. For example, the product , $\frac{2}{3} \times \frac{1}{2}$, may be determined using the following steps.

1. Begin with a rectangular piece of paper.
2. Fold the rectangle into two equal parts horizontally. Shade one of them using a pencil.
3. Now fold the rectangle into three equal parts vertically. Shade two of them using a pen.
4. The rectangle is now divided into sixths. Two of the sixths have been shaded twice.

   Therefore, the product is $\frac{2}{6}$ which can be rewritten as $\frac{1}{3}$.

Use the array model to determine the following products. Sketch your work.

1. $\frac{1}{2} \times \frac{3}{4}$

2. $\frac{1}{4} \times \frac{2}{3}$

3. $\frac{3}{8} \times \frac{1}{2}$

4. $\frac{2}{3} \times 4$

5. $\frac{2}{3} \times 2\frac{1}{2}$

ACTIVITY 6.3C - A DIVISION MODEL

Materials _required_: _Centimeter rods included with the textbook_

Centimeter rods may used to demonstrate the division of rational numbers. The quotient is determined by answering one of two questions.

      CASE ONE: The divisor is the smaller number. Example: $\frac{2}{3} \div \frac{1}{6}$.

          ASK, "HOW MANY sixths are there in two thirds?"

          SOLUTION: Let the J and B rods together represent one whole.

                Two D rods represent two thirds. The B rod represents one sixth.

                Compare the D and B rods. HOW MANY B rods are equal in length

                to the two D rods? Since the answer to this question is four, the

                quotient is four.

      CASE TWO: The divisor is the larger number. Example: $\frac{1}{6} \div \frac{2}{3}$.

          ASK, "HOW MUCH of two thirds is one sixth?"

          SOLUTION: Let the J and B rods together represent one whole.

                The B rod represents one sixth. Two D rods represent two thirds.

                Compare the B and D rods. What fraction of the two D rods is

                represented by the B rod? Since four B rods are equal in length to

                to the two D rods, the B rod is one fourth of the two D rods.

                Therefore, the quotient is $\frac{1}{4}$.

Determine the following quotients using your centimeter rods. Explain your work. Use the J and B rods together as one whole.

1. $\frac{1}{2} \div \frac{2}{3}$

2. $\frac{3}{4} \div \frac{1}{2}$

3. $\dfrac{7}{12} \div \dfrac{5}{6}$

4. $\dfrac{1}{6} \div \dfrac{3}{4}$

5. $\dfrac{5}{12} \div \dfrac{1}{4}$

6. $1 \div \dfrac{3}{4}$

7. $2 \div \dfrac{2}{3}$

8. $\dfrac{2}{3} \div 2$

9. $1\dfrac{1}{3} \div \dfrac{1}{12}$

ACTIVITY 6.3D-MULTIPLICATION & DIVISION PROPERTIES

Materials required: _Centimeter rods included with textbook_

1. Is the set of rational numbers commutative for multiplication? Does $\frac{1}{2} \times \frac{2}{3} = \frac{2}{3} \times \frac{1}{2}$?
    Explain your answer using each of the following models.

    a. Array model

    b. Centimeter rods

2. Is the set of rational numbers commutative for division? Does $\frac{2}{3} \div \frac{1}{4} = \frac{1}{4} \div \frac{2}{3}$?
    Use centimeter rods to justify your answer.

3. Is the set of rational numbers associative for multiplication?
    Does $\frac{1}{4} \times (\frac{1}{3} \times 2) = (\frac{1}{4} \times \frac{1}{3}) \times 2$? Explain your answer using each model.

    a. Array model

    b. Centimeter rods

6.3 CHECK YOUR UNDERSTANDING

1. Determine the product, $5 \times \frac{1}{4}$, using centimeter rods.

2. Determine the product, $3 \times 1\frac{1}{2}$, using centimeter rods.

3. Determine the product, $\frac{1}{3} \times \frac{1}{3}$, using an array model.

4. Determine the quotient, $\frac{1}{2} \div \frac{3}{4}$, using centimeter rods.

5. Determine the quotient, $\frac{5}{6} \div \frac{1}{3}$, using centimeter rods.

6. Is multiplication of rational numbers commutative? Explain using either an array or centimeter rods.

ACTIVITY 6.4A-IDENTIFYING & COMPARING ATTRIBUTES

Materials required: *Any two objects*

1. Select two objects from your environment.

   Object A:

   Object B:

2. Name five attributes which both objects A and B exhibit.

   a.

   b.

   c.

   d.

   e.

3. For each attribute in #2, give one non-standard(preferably original) unit for measuring it.

   a.

   b.

   c.

   d.

   e.

4. Compare the two objects using one of the attributes in #3.

5. For each attribute in #2, give one standard unit of measurement.

   a.

   b.

   c.

   d.

   e.

6. Compare the two objects using one of the4 attributes and the standard unit you suggested in #3.

ACTIVITY 6.4B - RATIO & PROPORTION

Materials required: _Tangrams from the Appendix_

I. Ratio

    A. Compare the number of triangles to the number of quadrilaterals. Write the comparison as a fraction. $\quad \dfrac{\text{number of triangles}}{\text{number of quadrilaterals}} = \underline{\hspace{2cm}}$

    B. There are five other comparisons, or ratios, which may be written using the tangrams.

        1. $\dfrac{\text{number of quadrilaterals}}{\text{total tangrams}} = \underline{\hspace{2cm}}$

        2. $\dfrac{\text{number of quadrilaterals}}{\text{number of triangles}} = \underline{\hspace{2cm}}$

        3. $\dfrac{\text{total tangrams}}{\text{number of quadrilaterals}} = \underline{\hspace{2cm}}$

        4. $\dfrac{\text{number of triangles}}{\text{total tangrams}} = \underline{\hspace{2cm}}$

        5. $\dfrac{\text{total tangrams}}{\text{number of triangles}} = \underline{\hspace{2cm}}$

    C. Write six ratios using the number of pieces which contain at least one right angle and the number of pieces which contain no right angles.

        1.

        2.

        3.

        4.

        5.

        6.

II. Proportion

Form a group of four with three of your classmates.

A. We increase the number of tangrams, but we keep the ratio of triangles to quadrilaterals constant.

1. Situation 1:

    a. How many group members must add their sets to yours to reach a total of fifteen triangles?

    b. How many quadrilaterals do you have then?

    c. Write a ratio which compares the number of triangles to the number of quadrilaterals.

    d. Is this equal to the ratio in Part IA? Explain.

    e. How many tangrams do we have?

    f. Write one ratio using the number in part e.

2. Situation 2:

    a. How many group members must add their sets to yours to reach a total of eight quadrilaterals?

    b. How many triangles do you have then?

    c. Write a ratio which compares the number of quadrilaterals to the number of triangles.

    d. Is this equal to the ratio in #2 of Part IB? Explain.

    e. How many tangrams do we have now?

    f. Write one ratio using the number in part e.

B. We increase the number of tangrams, but we preserve the ratio of pieces which contain at least one right angle to the number of pieces which contain no right angles.

1. Situation 1: Devise a plan for determining the number of tangram pieces you will have if you increase the number of pieces which contain no right angles to four.

2. Situation 2: Devise a plan for determining the number of pieces with at least one right angle if you know you have 21 tangrams.

ACTIVITY 6.4C - SCALE DRAWINGS

Materials required: *Ruler marked with inches*

Make a scale drawing of your room.

Let $\frac{1}{4}$ inch represent one foot.

Be certain to show:

1. The room dimensions.
2. The locations of the windows.
3. The location of the door.
4. The location and the approximate dimensions of the furniture.

6.4 CHECK YOUR UNDERSTANDING

Given the following collection of triangles and quadrilaterals:

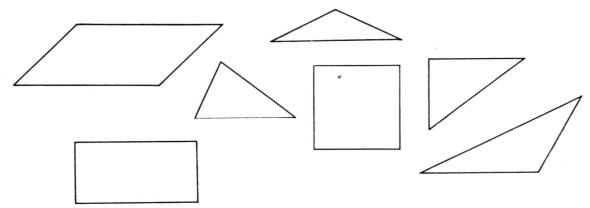

1. Write six ratios using this collection.

   a.   d.

   b.   e.

   c.   f.

2. We keep the ratio of the number of triangles to the number of quadrilaterals constant, but increase the number in the collection to 42. How many triangles will we have? Explain.

3. We keep the ratio of the number of triangles to the number of quadrilaterals constant, but we increase the number of quadrilaterals to 27. How many triangles will we have? Explain.

4. Suppose that we are told that in another collection the ratio of triangles to quadrilaterals is $\frac{3}{5}$. Can we determine the number of pieces in the collection? Explain.

7: DECIMALS AND PERCENTS

Introduction: The ideas developed in this section are particularly important for future elementary teachers. Many adults today appear to be lacking in facility with percents and the Metric System. These activities should give confidence in these crucial areas.

1. Using the sports pages of a recent newspaper, describe the way in which the use of decimals allows us to compare the performances of two individual athletes. Attach the newspaper clipping, or a copy of it, in the space provided below.

2. Find one application of percents greater than 100%. Explain the meaning of the magnitude of the percent and discuss the appropriateness of its use. The application may be from an advertisement, article, textbook, or conversation.

3. Will it cost the consumer less if the sales tax is computed before the discount is taken or if the discount is computed before the sales tax is computed? Explain your answer.

4. Top of the Tower Restaurant is offering an Early Bird Special. Each customer who arrives before 5:00 PM is given a 10% discount. Erika saves $1.10 and Angel saves $1.75. Why did Angel save more than Erika? Explain thoroughly.

5. Name one object from your environment which is normally measured using millimeters, centimeters, or meters. Estimate its dimensions using inches, feet, or yards.

6. Find three measurements which are given using scientific notation. List them below.

7. a. Estimate $\sqrt{19}$.

   b. Using your calculator enter 19 and press the square root button. Does it appear to terminate or repeat?

   c. Now push the x^2 button or the multiplication sign and equals. Did you get 19 as a result? Explain.

ACTIVITY 7.1 - DECIMAL REPRESENTATIONS

Materials required: **Base ten pieces from the Appendix**

Assume that the large block represents one whole.

I. Complete the following table.

	FLAT	LONG	UNIT CUBE
Number needed to form block			
Fractional representation			
Decimal representation			

II. Explain how you could represent each of the following using your pieces.

    A.   1.34

    B.   .654

    C.   .003

    D.   .203

III. Compare .305 and .350 using your pieces. Which is greater? Why?

IV. Let a flat represent one whole.
    A. Because there are _ _ _ _ _ longs in one flat, each long will represent the decimal _ _ _ _ _ _.
    B. Because there are _ _ _ _ _ unit cubes in one flat, each unit cube will represent _ _ _ _ _ _ _.
    C. Because there are _ _ _ _ _ flats in the large cube, the cube represents the decimal _ _ _ _ _ _ _.

149

ACTIVITY 7.2A - ADDITION OF DECIMALS

Materials required: Base ten pieces from the Appendix

EXAMPLE: We may add .324 and .708 using the base ten pieces. We will let one block represent one whole. We add three flats, two longs, and four unit cubes to seven flats and eight unit cubes. The result is ten flats, two longs, and 12 unit cubes. Because we are in base ten we must trade the ten flats for a large block and ten of the unit cubes for a long. We now have one block, three longs, and two unit cubes. Because the block represents one whole, the number may be written as 1.032.

Explain how one could add each of the following decimals using the base ten pieces. Let the block represent one whole.

A.   .068 + .135

B.   .007 + .994

C.   .907 + .36

D.   .37 + .752

ACTIVITY 7.2B - SUBTRACTION OF DECIMALS

Materials required: Base ten pieces from the Appendix

EXAMPLE: We may also demonstrate the difference .709 − .386 using the base ten pieces. If we let one block represent one whole, then we can represent .709 using seven flats and nine unit cubes. We must remove three flats, eight longs, and six unit cubes. We remove the six unit cubes and are left with three unit cubes. Next we must remove eight longs. We have no longs. One of the flats is traded for ten longs and the eight longs are subtracted. Three flats are now removed from the remaining six flats. The result is three flats, two longs, and three unit cubes, or .323.

Describe the way in which to perform the following subtractions using your base ten pieces. Let the block represent one whole.

A.   .628 − .435

B.   .853 − .568

C.   .82 − .004

D.   1.003 − .007

151

ACTIVITY 7.3 - PERCENTS USING A GEOBOARD

Materials required: Laminated geoboard included with textbook, geoboard, or dot paper

1. Most geoboards consist of sixteen squares in a four by four array. If we let the sixteen squares represent one whole, what percent of the geoboard do they represent?

2. Consider the geoboard which is shown at the right.

   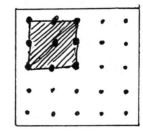

   a. How many squares are shaded?

   b. What fraction of the sixteen squares is shaded?

   c. Multiply this fraction by 100%. What is the result?

   d. What percent of the geoboard is shaded?

3. Represent the shaded portion of each geoboard below using a fraction and a percent.

   a.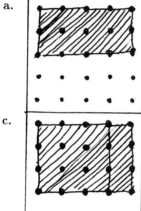

   b.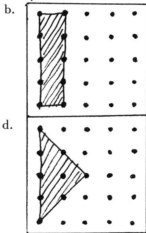

   c.

   d.

4. How could we represent $37 \frac{1}{2}$ % on a geoboard?

   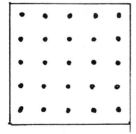

152

7.1-7.3 CHECK YOUR UNDERSTANDING

1. Which is greater .235 or .4? Refer to your base ten pieces in order to explain.

2. How could we demonstrate .608 + .397 using base ten pieces?

3. Explain how one could subtract .23 − .045 using base ten pieces.

4. How could 12 $\frac{1}{2}$ % be represented using a geoboard?

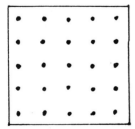

153

ACTIVITY 7.4 - METRIC LINEAR MEASUREMENT

Materials required: Scissors, paper, tape

I. The rectangle below is one decimeter long. It has been divided into ten equal parts called centimeters (cm) .

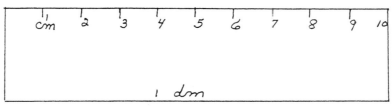

$$10 \text{ _____} = 1 \text{ _____}$$

$$1 \text{ _____} = .1 \text{ _____}$$

II. The rectangle below is also one decimeter long. It has been divided into one hundred equal parts called millimeters (mm) .

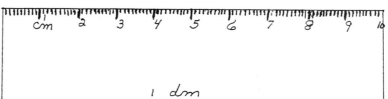

$$100 \text{ _____} = 1 \text{ decimeter}$$

$$10 \text{ _____} = 1 \text{ centimeter}$$

$$1 \text{ millimeter} = .1 \text{ _____}$$

$$1 \text{ millimeter} = .01 \text{ _____}$$

155

III. A. Trace the rectangle in Part II and cut out your copy. Be sure to trace the markings.

   B. Use this rectangle as a pattern to cut out nine additional rectangles.

   C. Tape the ten rectangles together to form a strip which is one meter in length.

   1 meter = 1000 _ _ _ _ _ _ _ _ _   .1 meter = 1 _ _ _ _ _ _ _ _ _ _

   1 meter = 100 _ _ _ _ _ _ _ _ _   .01 meter = 1 _ _ _ _ _ _ _ _ _ _

   1 meter = 10 _ _ _ _ _ _ _ _ _   .001 meter = 1 _ _ _ _ _ _ _ _ _

IV. A. Compare one centimeter to the width of your little finger.

   B. Compare your meter strip to the distance from the floor to the doorknob.

   C. Turn your head and look to the left. Raise your right arm at your side so that it is parallel to the floor. Compare your meter strip with the distance from your nose to the fingers on your right hand.

   D. Using a pencil, mark your height on a wall or door jam. Measure the distance from the floor to the mark in centimeters.

   E. Measure the length of your thumb in millimeters.

V. Estimate the following and check using your meter strip.

	ESTIMATE	ACTUAL
1. length of your pen		
2. length of the room		
3. width of your desk		

7.5A IRRATIONALS ON THE GEOBOARD

Materials *required*: *Geoboard or dot paper, graph paper, compass, straightedge*

Pythagorean Theorem: $a^2 + b^2 = c^2$, where c is the hypotenuse of a right triangle, and a and b are the other two sides, or legs.

1. Form a right triangle on your geoboard with legs of lengths 1 and 2. Give the exact value of the length of the hypotenuse. Is this number rational or is it irrational?

   SKETCH:

2. Form a right triangle on your geoboard with legs of lengths 3 and 2. Give the exact value of the length of the hypotenuse. Is this number rational or is it irrational?

   SKETCH:

3. Form a right triangle on your geoboard with legs of lengths 3 and 4. Give the exact value of the length of the hypotenuse. Is this number rational or is it irrational?

   SKETCH:

4. What conclusions can you draw from your work in #1 - 3?

5. If we wanted to construct a line segment with length equal to $\sqrt{13}$, could we use a geoboard? Show how this could be accomplished using graph paper.

6. Show how your would construct a line segment with length exactly equal to $\sqrt{74}$. Use graph paper.

7. Construct a line segment with length exactly equal to $\sqrt{7}$ using a ruler or a straightedge and a compass.

8. Show how you could construct a line segment with length exactly equal to $\sqrt{21}$ using a ruler or a straightedge and a compass.

ACTIVITY 7.5B - THE REAL NUMBERS

Materials required: Scissors, string

1. Cut out each of the set names below.

2. Make a Venn Diagram using string circles and your set names in order to show the relationships among the various types of numbers.. Make certain that you respect such notions as disjoint, intersection, and subset.

3. Cut out the squares. Each square is labeled with a number.

4. Place the squares appropriately within the Venn diagram.

Irrationals	$\sqrt{4}$	$\sqrt{7}$
Whole Numbers	$-\frac{4}{5}$	$\frac{9}{3}$
Integers	17	1.010010001...
Counting Numbers	$\frac{1}{3}$	$9.\overline{45}$
Rationals	0	19.5
Reals	4	-2

7.4 & 7.5 CHECK YOUR UNDERSTANDING

1. Complete.

   a.   1 decimeter = _____ millimeters

   b.   1 millimeter = _____ meters

   c.   1 meter   = _____ decimeters

2. Name the most appropriate unit of measurement (cm, mm, m) for each of the following lengths.

   a.   height of your bed

   b.   width of your hand

   c.   thickness of a pencil

3. Construct a line segment with length exactly equal to $\sqrt{83}$ using a ruler or a straightedge and a compass.

4. True of false?
   a. {irrationals} $\subset$ {reals}
   b. {rationals} $\cap$ {irrationals} = { }
   c. 0 $\in$ {whole numbers}

8: USING STATISTICS

Introduction: Daily we encounter conclusions and recommendations which are supposedly based upon statistical principles.   We must be aware of the way in which these are formulated in order to make sound decisions in our lives.   The activities in this unit will encourage this awareness.

I. DIRECTIONS: Find one EXAMPLE of each of the following graphs and attach it, or a copy of it, in the space provided below.   State one CONCLUSION that can be drawn from the information in each graph.

1. PICTOGRAPH

2. BAR GRAPH

3. LINE GRAPH

4. CIRCLE GRAPH

II. Do any of your graphs present the data in such a way that the conclusion appears more dramatic than it actually is? Explain.

ACTIVITY 8.1A - STEM AND LEAF PLOTS

I. Complete the following tests.

    A. Test 1: Presidents

        1. Time limit: 5 minutes

        2. Test: On a sheet of paper list the names of the men who served as president before William Howard Taft. (Hint: Taft served from 1909 to 1913.)

        3. Score: Check your answers with those at the end of this section. Give yourself four points for each correct response. Write your score on a slip of paper. Your instructor will collect them and write them on the board. (Do not write your name on the paper.)

    B. Test 2: Capitals

        1. Time limit: 5 minutes

        2. Test: On a sheet of paper list the names of the fifty state capitals. You need not name the corresponding states.

        3. Score: Check your answers with those at the end of this section. Give yourself two points for each correct response. Write your score on a slip of paper. Your instructor will collect them. (Do not write your name on the paper.)

II. Your instructor will list the scores for your class on the board. List the scores in order from least to greatest in the space provided below.

    Presidents:

    Capitals:

165

III. Compare the two sets of scores by making a two-sided stem and leaf plot. Put the scores for Test 1 on the left side of the stem and the scores for Test 2 on the right side of the stem. Remember that the scores must be in ascending order outward from the stem. The numbers on the stem represent the tens digits of the scores.

Test 1: Presidents   Test 2: Capitals

```
                    |  0  |
                    |  1  |
                    |  2  |
```

IV. What conclusions can you draw from the stem and leaf plot?

ACTIVITY 8.1B - THE HISTOGRAM

Materials required: Data from Activity 8.1A

I. GROUPING THE DATA

    A. If n is the number of scores, then n = _____.

    B. If k is the smallest number such that $2^k \geq n$, then k = _____.

        Example: Suppose n = 45.

$$2^k \geq 45$$

        The <u>smallest</u> value for k which would make this true is 6.

    C. To find the length of the intervals, divide the range by k, $\dfrac{\text{largest score } - \text{ smallest score}}{k}$.

        Each interval will include _____ units.

    D. Make a frequency table of _____ (k) intervals of _____ units each.

Test 1: Presidents			Test 2: Capitals	
Interval	Frequency		Interval	Frequency

167

II. Make a histogram for each set of data.

A. PRESIDENTS

B. CAPITALS

III. What conclusions can you draw from the histograms above?

ACTIVITY 8.1C - THE BAR GRAPH

Materials required: *Birthdays from Activity 4.4B.*

I. Make a bar graph using the data from Activity 4.4B. Let each division on the vertical scale represent one person. (If you need to extend the scale, trace the axes onto another sheet of paper and add more units of the same size.)

II. What conclusions can you draw from Part I?

III. Make another bar graph using the data from Activity 4.4B. Let each division on the vertical scale represent 5 people. (If you need to extend the axes, trace the graph onto another sheet of paper and add more units of the same size.)

IV. Does the graph give you the same impression as the one in Part I? Explain.

ACTIVITY 8.1D - THE LINE GRAPH

Materials required: String, metric tape measure from the Appendix, or string and centimeter ruler

I. With the help of a partner make the following measurements in centimeters.

> Your arm span: _ _ _ _ _ _ _ cm

> Your height: _ _ _ _ _ _ _ cm

II. Write the ordered pair, (height,arm span). _ _ _ _ _ _ _ _

> Your instructor will gather all of the ordered pairs for your class.
> List the ordered pairs for your class below.

III. Plot the ordered pairs using the axes below. Let each division of the vertical scale represent 30 cm. (If you need to extend the vertical axis, you may trace the axes onto another sheet of paper and add more divisions of the same size.)

IV. What is the trend in the data?

V. Use your string to find a line of best fit. Position the string so that it seems to pass through the middle of the plotted points and to follow the trend of the data. Estimate the slope of the line of best fit. _____

VI. What conclusions can you draw from the graph?

VII. Make two more line graphs using the axes below. Let the vertical scale in A begin with 50 cm and let the vertical scale in B begin with 100 cm. Let each division of the vertical scales represent 30 cm.

A.

B.

VIII. Do the graphs in Part VII leave a different impression than that in Part III? Explain.

Graph VII - A:

Graph VII - B:

IX. What conclusions can be drawn from your work in Part VIII?

ACTIVITY 8.1E - THE CIRCLE GRAPH

Materials required: *Data from Activity 4.4B, protractor, and compass*

I. In Activity 4.4B we determined the days of the week on which we were born. Complete the following chart using the days on which your classmates were born.

DAY OF WEEK	NUMBER OF CLASS MEMBERS	FRACTION OF CLASS	PERCENT OF CLASS	MEASURE OF CENTRAL ANGLE
SUNDAY				
MONDAY				
TUESDAY				
WEDNESDAY				
THURSDAY				
FRIDAY				
SATURDAY				

II. Make a circle graph for the data.

III. What conclusions can you draw from the circle graph above?

ACTIVITY 8.1F - THE PICTURE GRAPH

Materials required: *Data from Activity 8.1E*

I. Select an appropriate symbol to represent each person's birthday or seven different symbols to represent birthdays on different days of the week.

   SUNDAY:

   MONDAY:

   TUESDAY:

   WEDNESDAY:

   THURSDAY:

   FRIDAY:

   SATURDAY:

II. Make a picture graph of the data from Activity 8.1E.

III. Suppose there had been 300 people in your class.

How would you use your chosen symbol(or symbols) to represent each of the following?

A. Fifty-five people born on Wednesday, if each symbol represents 10 people.

B. One hundred seventy-three people born on Friday, if each symbol represents 25 people.

C. Two hundred four people born on Sunday, if each symbol represents 30 people.

IV. Suppose there had been 550 people in your class.

How would you use your chosen symbol(or symbols) to represent each of the following?

A. Thirty-three people born on Tuesday, if each symbol represents 50 people.

B. One hundred forty-seven people born on Monday, if each symbol represents 25 people.

C. Two hundred sixty-eight people born on Saturday, if each symbol represents 75 people.

8.1 CHECK YOUR UNDERSTANDING

1. a. Make a two-sided stem and leaf plot to compare the following math test scores.

   EXAM 1:   55, 56, 59, 60, 63, 70, 75, 83, 88, 89

   EXAM 2:   60, 60, 62, 63, 65, 68, 72, 90, 90, 92

   b. On which test did the class perform better?

2. Bar graphs are especially good for showing _____.

3. Line graphs are especially good for showing _____.

4. Give one way in which bar graphs may be deceptive.

5. Give one way in which line graphs may be deceptive.

6. If 35% of a class selects the turtle as their favorite reptile, how many degrees will be in the central angle of that sector in the circle graph?

7. In a group of 600 students, 100 select red as their favorite color. If we make a picture graph in which each symbol represents 50 people, how many symbols will we need to represent the number of people who like red best?

ACTIVITY 8.2A - MEASURES OF CENTRAL TENDENCY

Materials required: *Data from Activity 8.1A*

I. Use the test scores from Activity 8.1A to complete the following charts.

PRESIDENTS: CAPITALS:

Mean: Mean:

Median: Median:

Mode: Mode:

II. Make comparisons.

A. Compare the three measures for each set of data.(Hint: Graph the three measures for each set of data on a separate number line.)

B. For which set of data are the measures closest? What is the significance of this?

C. What conclusions can you draw from these comparisons?

ACTIVITY 8.2B - MEASURES OF DISPERSION

Materials required: ***Data from Activity 8.1A***

I. Determine the Variance and Standard Deviation for each set of data in Activity 8.1A.

PRESIDENTS: Mean($\overline{A}$) :

Variance:

Standard Deviation:

Score(A_i)	$A_i - \overline{A}$	$(A_i - \overline{A})^2$

CAPITALS: Mean($\overline{A}$) :

Variance:

Standard Deviation:

Score(A_i)	$A_i - \overline{A}$	$(A_i - \overline{A})^2$

II. Discuss the dispersion of each set of data. What is the significance of the standard deviation? Compare the two sets of data using their standard deviations.

III. Determine the z-scores for each set of data in Activity 8.1A.

PRESENTS:

Score(A_i)	z_i

CAPITALS:

Score(A_i)	z_i

IV. Select a score that appears in both tables above. What do the corresponding z-scores tell you?

ACTIVITY 8.2C - THE BOX & WHISKER PLOT

Materials required: *Data from Activity 8.1A*

I. Complete each chart using the data from Activity 8.1A.

PRESIDENTS: CAPITALS:

Q_1(Lower Quartile): Q_1(Upper Quartile):

Q_2(Median): Q_2(Median):

Q_3(Upper Quartile): Q_3(Upper Quartile):

Lowest Extreme Value: Lowest Extreme Value:

Highest Extreme Value: Highest Extreme Value:

II. Make a box and whisker plot for each set of data.

PRESIDENTS:

CAPITALS:

III. Compare the performance of the class on the two tests using your box and whisker plots.

NAME: _____ DATE _____

8.2 CHECK YOUR UNDERSTANDING

Given the following scores of the same class on two different exams:

Math Test: 58, 70, 70, 86, 88, 92, 96

History Test: 55, 55, 67, 70, 71, 80, 99

1. a. Determine the measures of central tendency for each set of data.

MATH TEST   HISTORY TEST

Mean: Mean:

Median: Median:

Mode: Mode:

b. Compare the performance of the class on the two exams.

2. a. Determine the variance and standard deviation to nearest tenth for each set of data.

MATH TEST   HISTORY TEST

Variance: Variance:

Standard deviation: Standard deviation:

b. Compare the performance of the class on the two exams using the standard deviation.

c. A student scores 70 on both exams. Compare his performance using the z-scores.

4. a. Complete each chart and make a box and whisker plot for each set of test scores.

MATH TEST

Q_1:

Q_2:

Q_3:

Lowest extreme score:

Highest extreme score:

HISTORY TEST

Q_1:

Q_2:

Q_3:

Lowest extreme score:

Highest extreme score:

MATH TEST

HISTORY TEST

b. On which test did the class perform better? Explain.

TEST 1:   PRESIDENTS OF THE UNITED STATES BEFORE WILLIAM HOWARD TAFT

1. George Washington
2. John Adams
3. Thomas Jefferson
4. James Madison
5. John Quincy Adams
6. Andrew Jackson
7. Martin Van Buren
8. William Henry Harrison
9. John Tyler
10. James K. Polk
11. Zachary Taylor
12. Millard Fillmore
13. Franklin Pierce
14. James Buchanan
15. Abraham Lincoln
16. Andrew Johnson
17. Ulysses S. Grant
18. Rutherford B. Hayes
19. James A. Garfield
21. Chester A. Arthur
22. Grover Cleveland
23. Benjamin Harrison
24. William McKinley
25. Theodore Roosevelt

TEST 2:   THE STATE CAPITALS

Albany (New York)

Annapolis (Maryland)

Atlanta (Georgia)

Augusta (Maine)

Austin (Texas)

Baton Rouge (Louisiana)

Bismarck (North Dakota)

Boise (Idaho)

Boston (Massachusetts)

Carson City (Nevada)

Charleston (West Virginia)

Cheyenne (Wyoming)

Columbia (South Carolina)

Columbus (Ohio)

Concord (New Hampshire)

Denver (Colorado)

Des Moines (Iowa)

Dover (Delaware)

Frankfort (Kentucky)

Harrisburg (Pennsylvania)

Hartford (Connecticut)

Helena (Montana)

Honolulu (Hawaii)

Indianapolis (Indiana)

Jackson (Mississippi)

Jefferson City (Missouri)

Juneau (Alaska)

Lansing (Michigan)

Little Rock (Arkansas)

Lincoln (Nebraska)

Madison (Wisconsin)

Montgomery (Alabama)

Montpelier (Vermont)

Nashville (Tennessee)

Oklahoma City (Oklahoma)

Olympia (Washington)

Phoenix (Arizona)

Pierre (South Dakota)

Providence (Rhode Island)

Raleigh (North Carolina)

Richmond (Virginia)

Sacramento (California)

St. Paul (Minnesota)

Salem (Oregon)

Salt Lake City (Utah)

Sante Fe (New Mexico)

Springfield (Illinois)

Tallahassee (Florida)

Topeka (Kansas)

Trenton (New Jersey)

9: PROBABILITY

Introduction: A study of probability will give us the ability to make and to better understand predictions. Give an actual example of each of the following and explain one way in which the laws of probability are involved in the outcome of each situation.

1. WEATHER FORECAST:

2. BIRTH ANNOUNCEMENT:

3. LOTTERY DRAWING OR PROMOTIONAL GIVE-AWAY:

4. OTHER SITUATION:

9.1A - EXPERIMENTAL PROBABILITY

Materials required: Ten thumbtacks, paper cup(or empty can)

I. Thumbtack Experiment

    A. List the two possible outcomes when a thumbtack is tossed.

    B. In 100 tosses of the thumbtack, do you expect each of these events to occur 50 times? Explain.

    C. Because we cannot assume that the outcomes are equally likely, we must conduct an experiment to determine the probability of the thumbtack landing point up, P(point up).

       1. Put ten thumbtacks in a paper cup.

       2. After shaking the cup, pour out the thumbtacks.

       3. Count the number of thumbtacks which land point up.

       4. Record the number in the chart below.

       5. Repeat the experiment another nine times. (For a total of ten times.)

Toss Number	1	2	3	4	5	6	7	8	9	10
Number landing point up										

       Total landing point up: _ _ _ _ _ _ _

      6. $\text{P(point up)} = \dfrac{\text{number landing point up}}{\text{total thumbtacks tossed}} = \dfrac{}{100}$

      7. Why is the denominator 100?

8. P(point down) $= \dfrac{\text{number landing point down}}{\text{total thumbtacks tossed}} = \dfrac{\quad}{100}$

9. P(point up) + P(point down) $= \dfrac{\quad}{100} + \dfrac{\quad}{100} = \dfrac{\quad}{100}$ or ____

These are _____events.

II. Cup (or empty can) Experiment

A. List the three outcomes which are possible when a cup(or empty can) is tossed.

B. In thirty tosses of the cup(or can), do you expect each of the outcomes to occur ten times? Explain.

C. Because we cannot assume that the events are equally likely, we must conduct an experiment to determine the probability that the cup(or can) lands on its side, P(side).

1. Toss the cup(can).
2. Record the outcome below.
3. Repeat the experiment twenty-nine more times. (For a total of thirty times.)

Outcome	Tally	Frequency

4. Determine the probability of each outcome.

   P(   ) $= \dfrac{\quad}{30}$ P(   ) $= \dfrac{\quad}{30}$ P(   ) $= \dfrac{\quad}{30}$

5. Determine the sum of the probabilities in number 4. What is the significance of this?

ACTIVITY 9.1B - THE UNION OF TWO EVENTS

Materials required: _Attribute blocks from the Appendix_

An attribute block is chosen at random from the collection provided.

I. As in set theory, A ∪ B represents A or B. Remember that in mathematics "or" is inclusive. This means A, B, or BOTH. Explain the meaning of each of the following.

      A. P(yellow block or blue block)

      B. P(red block or triangle)

II. Determine the following.

      A. P(yellow block) =

      B. P(blue block) =

      C. P(blue _and_ yellow block) =

III. Does P(yellow block or blue block) = P(yellow block) + P(blue block)? Explain.

IV. Determine the following.

      A. P(red block) =

      B. P(triangle) =

      C. P(red triangle) =

V. Does P(red block or triangle) = P(red block) + P(triangle)? Explain.

VI. Summary. P(A ∪ B) = P(A or B) =

191

ACTIVITY 9.1C-THEORETICAL VS. EXPERIMENTAL PROBABILITY

Materials required: _A pair of dice_

I. Theoretical Probability

    A. Show the sample space of all possible sums when two dice are rolled.

    B. There are _____different sums.

       Many of the sums are the result of several different rolls of the dice. A sum of six may result from a 1 on the first die and a five on the second die, a two on the first die and a four on the second die, a three on each die, a four on the first die and a two on the second die, and a five on the first die and a one on the second die. All of these outcomes are equally likely.

       Therefore, there are really _____ different ways in which the dice may land.

    C. If the dice are rolled thirty-six times, how many times would we expect each of the following sums to occur?

Sum	2	3	4	5	6	7	8	9	10	11	12
Expected frequency in 36 trials											

    D. If the dice are rolled seventy-two times, how many times would we expect a sum of 8?

II. Experimental Probability

    A. Roll the dice.

    B. Add the results.

    C. Record the sum below.

    D. Repeat the experiment thirty-five more times. (For a total of thirty-six times.)

Sum	Tally	Frequency	Average Frequency
2			
3			
4			
5			
6			
7			
8			
9			
10			
11			
12			

    E. Compare the frequency of each outcome in Part II D with its expected frequency in Part I D. Are they identical? Why do you think this is the case?

    F. Determine the sum of your frequencies in Part II D and those of three of your classmates. Divide each sum by four to determine the average. Record the averages in the table in Part II D. Compare these averages with the expected frequencies in Part I C. Are they closer than your outcomes in Part II D? Why do you think this is the case? Predict the result if we use the results from the entire class in our averages.

    G. Summary. As the number of trials increases, the _ _ _ _ _ _ _ _ _ _ probability approaches the _ _ _ _ _ _ _ _ _ _ probability.

9.1 CHECK YOUR UNDERSTANDING

1. Give the formula for determining theoretical probability.

2. a. Describe one situation in which we cannot use the formula in number one.

   b. Explain why the formula cannot be used.

   c. How will we determine the probability?

3. a. If the P(A) is $\frac{3}{7}$, what is the P($\overline{A}$) ?

   b. What kind of events are these?

4. How is the theoretical probability related to the experimental probability for a situation in which the outcomes are equally likely?

5. When is the following statement true?
   $$P(A \cup B) = P(A) + P(B)$$

ACTIVITY 9.2A - CONDITIONAL PROBABILITY

Materials required: A die, a coin

I. A die is rolled and a coin is tossed simultaneously.

    A. Use a tree diagram to determine the sample space.

        ---------------------- TREE DIAGRAM ----------------------   --- SAMPLE SPACE ---

    B. Determine the probability of each outcome in the sample space.

    C. Determine the sum of the probabilities in Part B. Explain the significance.

II. In conditional probability we put a restriction on the sample space. This generally decreases the number of outcomes considered favorable.

    A. P(2 on the die, given that we tossed a head on the coin) means that we must now consider only those members of the sample space which include a head on the coin.

        1. List these outcomes below.

        2. Of the outcomes which you just listed, how many include a 2 on the die?

        3. Therefore, P(2 on the die, given that we tossed a head on the coin) =

197

B. P(head on the coin, given that we rolled an even number on the die) means that we must consider only those members of the sample space which include an even number on the die.

   1. List these outcomes below.

   2. Of the outcomes which you just listed, how many include a head on the coin?

   3. Therefore, P(head on the coin, given that we rolled an even number on the die) =

C. We may represent P(A, given that we know B) as P(A | B). B is the restriction which we place on our sample space. Determine the following probabilities.

   1. P(number less than five on the die | head on the coin)

   2. P(number greater than four on die | tail on the coin)

   3. P(head on the coin | number less than five on the die)

   4. P(tail on the coin | number greater than four on die)

D. Does P(A | B) = P(B | A)?

E. If A and B are independent, then P(A | B) = P(A). If A is the event, "tail on the coin," and B is the event, "four on the die," are the events independent? Explain.

ACTIVITY 9.2B - EXPECTED VALUE

I. <u>GAME 1</u>: Multiplying Dice

Each time you participate in this game, you must pay $4.00. You roll the dice. If their product is even then you must pay $2.00. If their product is odd, then you win $7.00.

    A. Complete the following chart in order to determine the sample space.

X	1	2	3	4	5	6
1						
2						
3						
4						
5						
6						

    B. Determine the following probabilities.

        1. P(odd product) =

        2. P(even product) =

    C. In order to determine your expected winnings, you may perform the following operation.

        7 [P(odd product)] $-$ 2 [P(even product)] $-$ 4 =

    D. Is it wise to play this game? Explain.

    E. Suppose we change the rules. You must now pay $5.00 to play. If the product is greater than or equal to twelve, then you win $18.00. If the product is less than twelve, then you pay $9.00. Determine your expected winnings.

II. <u>GAME</u> 2: Additional Dice

In this game we use the original rules for Game 1, but we find the sum of the dice instead of the product of the dice.

    A. Complete the following chart in order to determine the sample space.

+	1	2	3	4	5	6
1						
2						
3						
4						
5						
6						

    B. Determine the following probabilities.

        1. P(odd sum) =

        2. P(even sum) =

    C. Determine your expected winnings for this game.

    D. Is it wise to play this game? Explain.

    E. Suppose we change the rules. You will pay $1.50 to play. If the sum is greater than or equal to five, then you will win $2.00. If the sum is less than five, then you will pay $1.00. Determine your expected winnings.

NAME _____ DATE _____

9.2 CHECK YOUR UNDERSTANDING

1. Make a tree diagram to determine the sample space for an experiment in which a coin is tossed and the spinner at right is spun.

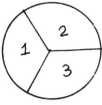

2. Determine P(the coin shows heads | the spinner is odd) .

3. Does P(A | B) = P(A) for the two events described in number 2? Are the events independent?

4. Rules of the game: You pay $2.00 to play. You roll a die. If the number is even, then you win $1.00. If the number is odd, then you lose $.50. What are your expected winnings?

ACTIVITY 9.3A - PASCAL'S TRIANGLE

Materials required: **Three coins**

I. List the sample space for each of the following numbers of coins and then complete the chart at right.

    A. One Coin

Sample Space

number of heads	0	1
probability		

    B. Two Coins

Sample Space

number of heads	0	1	2
probability			

    C. Three Coins

Sample Space

number of heads	0	1	2	3
probability				

    D. Complete the first ten rows of Pascal's Triangle.

$$1$$
$$1 \qquad 1$$
$$1 \qquad 2 \qquad 1$$

E. Compare the numbers in your charts in Parts A, B, and C with the rows of Pascal's Triangle. What do you suppose that the row, 1 3 3 1, represents? What is the significance of the second entry in each row?

F. Add the entries in each horizontal row. Compare those sums with the numbers of outcomes in the sample spaces. What is the sum of the row, 1   3   3   1? What do you think that sum represents?

G. Use Pascal's Triangle to determine the number of outcomes in the sample space when seven coins are tossed. (Hint: Use the row that has a second entry equal to seven.)

H. Write the entries in the row that you used in Part G. Give an interpretation of these entries.

I. Use Pascal's Triangle to determine the following probabilities.

    1. P(2 heads when 6 coins are tossed) =

    2. P(4 heads when 8 coins are tossed) =

    3. P(3 heads when 5 coins are tossed) =

J. How might you use Pascal's Triangle to determine the probability of getting 4 girls in a family of seven children? Explain.

II. Three Coin Experiment

A. Toss three coins.

B. Record the number of heads.

C. Repeat the experiment twenty-three times. (For a total of 24.)

Number of Heads	Tally	Frequency	$\dfrac{\text{Frequency}}{3}$
0			
1			
2			
3			

D. Compare the last column of your table with the frequencies in your table in Part I C.

E. Average the entries of your last column in Part I C with those of three other people in your class. Record the results below.

Number of Heads	0	1	2	3
Average Frequency				

F. Compare the average frequencies in Part II E with those in the last column of your table in Part I C.

G. Summary. As the number of trials increases, the _____ probability approaches the _____ probability.

205

ACTIVITY 9.3B - PERMUTATIONS VS. COMBINATIONS

I. Given the letters, ABCD, make a list of all the possible arrangements of those letters taken three at a time. (You may wish to use a tree diagram.)

A. Permutations of four letters taken three at a time.

1. .How many arrangements are there? These are the permutations of the letters.

2. Notice that the number of permutations, or arrangements, of the four letters taken three at a time could be determined using the following process.

   We have 4 choices for the first letter <u>and</u> three choices for the second <u>and</u> two choices for the third.

   Therefore, there are 4 × 3 × 2 , or 24, different arrangements, or permutations, of the given letters.

3. Had we been asked to determine the number of permutations of six letters taken four at a time, how could we avoid making a tree diagram?

B. Permutations of three letters taken three at a time.

    1. Group the permutations so that each group contains different arrangements of the same three letters. How many arrangments are in each group?

    2. We could have also determined the number of permutations of each group of three letters using the following process.

       There are three choices for the first letters <u>and</u> two for the second <u>and</u> one for the third. Therefore, there are $3 \times 2 \times 1$, or six, different arrangements, or permutations, of each group of three letters.

    3. Had we been asked to determine the number of permutations of four letters taken four at a time, how could we avoid grouping our outcomes?

C. Combinations of four letters taken three at a time.

    1. Determine the ratio, $\dfrac{\text{Permutations of 4 things taken 3 at a time}}{\text{Permutations of 3 things taken 3 at a time}}$. How does this compare to the number of groups?

    2. Therefore, to determine the number of groups, or combinations, of 7 letters taken 4 at a time, we could use the following ratio.

$$\frac{\text{Permutations of 7 things taken 4 at a time}}{\text{Permutations of 4 things taken 4 at a time}} = \frac{7 \times 6 \times 5 \times 4}{4 \times 3 \times 2 \times 1} = 35$$

This means that we can form 35 different combinations of four letters using the seven letters we are given.

D. Given the letters ABCDE.

   1. List all the permutations of the five letters taken two at a time. (You may need to use a tree diagram to organize your list.)

   2. How many permutations are there? How does this compare with 5 × 4?

   3. Group your permutations so that each group contains arrangements of the same two letters. How many permutations are there in each group? How does this compare with 2 × 1?

   4. $\dfrac{\text{Permutations of 5 things taken 2 at a time}}{\text{Permutations of 2 things taken 2 at a time}} = \underline{\hspace{2cm}} =$

   5. How does the result in number 4 compare with the number of different groups in number 1? This is the number of combinations of five letters taken two at a time.

E. Outline the ways in which you could determine the number of combinations of ten things taken four at a time.

   1. Using a list.

   2. Using a ratio.

9.3 CHECK YOUR UNDERSTANDING

1. List the first eight rows of Pascal's Triangle.

2. Use Pascal's Triangle to determine the following probabilities.

    a. P(2 heads on four coins) =

    b. P(3 heads on six coins) =

    c. P(6 tails on seven coins) =

3. Contrast "permutation" and "combination".

4. Given: ABCDEF

    a. Determine the number of permutations of the six letters taken four at a time.

    b. Determine the number of combinations of six letters taken four at a time.

10: SPATIAL VISUALIZATION

Introduction: Through the activities in this chapter we will improve our ability to identify and classify geometric shapes and solids.

I. Find at least one example of each of the following geometric shapes in your environment. Give the location in which you found it and at least three characteristics that it exhibits.

SHAPE   LOCATION   CHARACTERISTICS

A. Triangle

B. Parallelogram

C. Rectangle

D. Square

E. Pentagon

F. Hexagon

G. Octagon

II. Find at least one example of each of the following three-dimensional figures in your environment. Give the name of each object and at least three characteristics that it exhibits.

FIGURE   OBJECT   CHARACTERISTICS

A. Pyramid

B. Prism

C. Sphere

D. Hemisphere

E. Cone

F. Cylinder

ACTIVITY 10.1A - THE CIRCLE

Materials required: *Piece of string, thumbtack, posterboard (or cardboard)*

I. Defining a circle

    A. Attach one end of the string to the center of the posterboard using the thumbtack.

    B. Attach a pencil to the other end of the string.

    C. Stretching the string tightly, draw a circle around the thumbtack.

    D. Label the following.

        1. Interior

        2. Exterior

        3. Circle

    E. In your own words define a circle.

II. Describing a circle

    A. On the circle which you drew in Part I indicate an example of each of the following.

        1. Radius

        2. Chord

        3. Diameter

        4. Circumference

    B. In your own words describe the meaning of each of the words in Part A.

        1. Radius:

        2. Chord:

        3. Diameter:

        4. Circumference:

ACTIVITY 10.1B - CONVEX AND NONCONVEX POLYGONS

Materials required: *Geoboard*

A polygon is <u>convex</u> if the line segment connecting any two points in its interior lies entirely within the interior of the figure. In a <u>nonconvex</u> polygon, this principle is violated.

<u>Examples:</u>

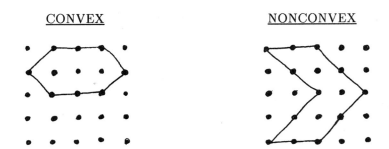

CONVEX   NONCONVEX

If possible, construct an example of each of the following polygons on your geoboard. Indicate which of the polygons are regular.

<u>POLYGON</u>	<u>CONVEX</u>	<u>NONCONVEX</u>
1. Triangle		
2. Quadrilateral		

214

POLYGON	CONVEX	NONCONVEX
3. Pentagon		
4. Hexagon		
5. Octagon		
6. Decagon		

215

ACTIVITY 10.1C - CHARACTERISTICS OF POLYGONS

I. GAME 1: WHAT'S MY POLYGON?

Number of players: Two or more

Materials required: Polygon cards

Rules of play:

1. Place the polygon cards face down on the table.

2. One player selects a card from the stack.

3. Beginning with the player on his right, the other players take turns asking questions, concerning characteristics of the polygon on the card, to which the player in #2 can only respond "yes" or "no."

4. Play continues in a counterclockwise direction until someone correctly identifies the polygon on the card.

5. The player in #2 is awarded one point for each question he is asked until the polygon is identified.

6. The player who correctly identifies the polygon is the next player to draw a card from the stack.

7. Play continues until all the cards in the stack have been used.

8. The player with the highest point total is the winner.

II. GAME 2: POLYGON TRANSFORMATIONS

Number of players: Two

Materials required: Polygon cards, geoboards

Rules of play:

1. Place the polygon cards face down on the table.

2. Each player selects a card and turns it face up on the table in front of him.

3. Each player must construct the polygon on his card on his geoboard.

4. Players then exchange cards.

5. Each player in turn must demonstrate the number of steps required to transform his polygon into that of his opponent. One step is defined as releasing from or wrapping around a specific nail.)

6. Each player scores one point for each step which he took in #5.

7. The player with the lowest number of points is the winner.

216

III. <u>GAME</u> <u>3</u>: CONTRAST AND COMPARISON

Number of players: Two or more

Materials required: Polygon cards, die

Rules of play:

1. Place the polygon cards face down on the table.

2. Each player rolls the die. Player with greatest number goes first.

3. First player rolls the die again and turns over the number of cards equal to the number on the die. He must then identify one characteristic which is common to all the shapes on the cards.

4. If the player's response in #3 is unacceptable to the other players, then he is awarded one point for each card.

5. Play continues with the next player on the right until there are insufficient cards for the number on the die.

6. The player with the smallest number of points is the winner.

ACTIVITY 10.1D - POLYGONS FROM TANGRAMS

Materials required: *Tangrams from the Appendix*

I. Identify each of your tangram pieces. Give the most precise classification. List them below.

II. Name five figures which can be formed using exactly two of your tangram pieces. Identify the pieces used, the figure formed, and whether the figure is convex. (You may use a sketch.)

Note: *Look for triangles, squares, rectangles, parallelograms, trapezoids, pentagons, hexagons, etc.*

FIGURE	TANGRAM PIECES	CONVEX?

III. Name five figures which can be formed using exactly three of your tangram pieces. Identify the pieces used, the figure formed, and whether the figure is convex. (You may use a sketch.)

FIGURE	TANGRAM PIECES	CONVEX?

IV. Name five figures which can be formed using exactly four of your tangram pieces. Identify the pieces used, the figure formed, and whether the figure is convex. (You may use a sketch.)

    <u>FIGURE</u>   <u>TANGRAM PIECES</u>   <u>CONVEX</u>?

V. Name five figures which can be formed using exactly five of your tangram pieces. Identify the pieces used, the figure formed, and whether the figure is convex. (You may use a sketch.)

    <u>FIGURE</u>   <u>TANGRAM PIECES</u>   <u>CONVEX</u>?

VI. Name five figures which can be formed using exactly six of your tangram pieces. Identify the pieces used, the figure formed, and whether the figure is convex. (You may use a sketch.)

<u>FIGURE</u>   <u>TANGRAM PIECES</u>   <u>CONVEX</u>?

VII. Name five figures which can be formed using all seven of your tangram pieces. Identify the figure formed and whether the figure is convex. (You may use a sketch.)

<u>FIGURE</u>   <u>TANGRAM PIECES</u>   <u>CONVEX</u>?

220

10.1 CHECK YOUR UNDERSTANDING

1. Define "circle" in your own words.

2. Give one example of a nonconvex polygon and explain why it is not convex.

3. Give the defining characteristics of each of the following.

   a. Square

   b. Parallelogram

   c. Trapezoid

   d. Octagon

   e. Dodecagon

4. What is meant by a *regular* polygon?

ACTIVITY 10.2A - MODELS OF REGULAR POLYHEDRA

Materials required: *Posterboard, hole punch, scissors, rubber bands*

INSTRUCTIONS: 1. Trace each of the following patterns.

2. Using the patterns, cut from posterboard the indicated number of polygons for each model.

3. Fold each piece along the dotted lines.

4. Use a hole punch at each corner.

5. Connect adjacent sides by putting their "insides" together and looping a rubber band around them. The rubber band should rest securely in holes that you punched at each corner.

I.   Tetrahedron: Make 4 equilateral triangles.

II. Hexahedron: Make 6 congruent squares.

III. Octahedron: Make 8 equilateral triangles.

IV. Dodecagon: Make 12 regular pentagons.

V.   Icosahedron: Make 20 equilateral triangles.

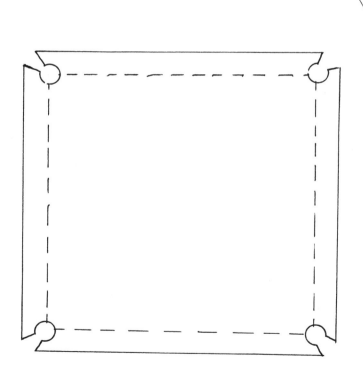

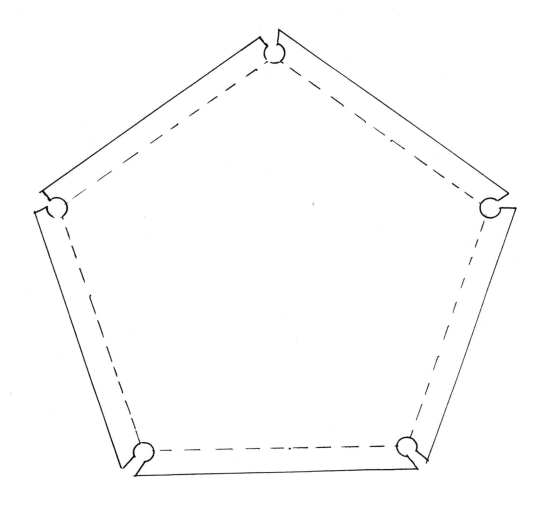

ACTIVITY 10.2B - EULER'S THEOREM

Materials required: *Models of prisms and pyramids from the Appendix, regular polyhedra models from Activity 10.2A, scissors, tape*

I. Assemble the models of the prisms and pyramids using tape.

II. Give some defining characteristics of the following solids.

    A. Prism   B. Pyramid

III. Can any of your regular poyhedra models be classified as prisms or pyramids? Explain.

IV. Complete the following chart by counting the number of vertices, edges, and faces of each of the models.

POLYHEDRON	NUMBER OF VERTICES	NUMBER OF FACES	NUMBER OF EDGES
A. Tetrahedron			
B. Hexahedron			
C. Octahedron			
D. Dodecahedron			

POLYHEDRON	NUMBER OF VERTICES	NUMBER OF FACES	NUMBER OF EDGES
E. Icosahedron			
F. Triangular Pyramid			
G. Square Pyramid			
H. Triangular Prism			
I. Square Prism			
J. Rectangular Prism			

V. Is there a relationship among the number of vertices, the number of faces, and the number of edges for polyhedra? Explain.

VI. Write the relationship in Part V as an equation. This is Euler's theorem.

VII. Use the formula to answer the following questions.

A. A polyhedron has 30 faces and 30 vertices. How many edges does it have?

B. A "polyhedron" has 4 vertices and 5 edges. Is this possible? Explain.

VIII. Does the formula hold for cones, cylinders, spheres, and hemispheres? Are they polyhedra? Explain.

ACTIVITY 10.2C - CROSS SECTIONS

Materials required: _Clay, plastic knife, wax paper (optional)_

I. Conic sections

    A. Using the clay form a cone. (Wax paper may facilitate cleanup.)

    B. Slice the cone parallel to the base using the knife. What plane figure appears at the place where the cut was made?

    C. Reshape the clay into a cone.

    D. With a single cut, slice both the apex and the base. What plane figure appears at the place where the cut was made?

    E. Reshape the clay into a cone.

    F. Slice through the cone without cutting the base or the apex and without your knife being parallel to the base. What plane figure appears at the place where the cut was made?

    G. Reshape the clay into a cone.

    H. Slice through the base of the cone without cutting the apex. Make your cut parallel to the side of the cone. What plane figure appears where the cut was made?

    I. Reshape the clay into a cone.

    J. With a partner, hold one cone above another cone so that the apex of your cone touches the apex of the other cone and the bases are parallel. Slice through both cones with a single straight cut. Do not slice the apex of either cone. What is the name of the resulting shape?

II. Cross sections of other three dimensional figures.

    A. Using your clay form each of the following figures.

    B. Slice each figure in various ways using the knife.

    C. Record the plane figures which result from each slice.

    D. Describe the way in which the slice was made for each cross section.

<u>THREE</u> <u>DIMENSIONAL</u> <u>FIGURE</u>   <u>POSSIBLE</u> <u>CROSS</u> <u>SECTIONS</u>

    1. Sphere

    2. Cylinder

    3. Cube

    4. Tetrahedron

10.2 CHECK YOUR UNDERSTANDING

1. State Euler's Theorem.

2. Use Euler's Theorem to answer the following questions.

   a. If a polyhedron has 28 vertices and 42 edges, how many faces does it have?

   b. Is it possible for a "polyhedron" to have 30 vertices and 31 edges? Why or why not?

3. List five cross sections of a cone.

4. Will any cross section of a cube produce an octagon? Explain.

11: MEASUREMENT

Introduction: The ability to measure the attributes of objects gives us a means of comparison and description.   The activities in this section will help us to better understand the attributes being measured and to have a greater facility with the instruments and the methods of measurement.

1. a. What are we actually measuring when we measure an angle?

    b. Without using a protractor, how could we convince the child that the following angles are equal in measure?

2. Find three uses of square units in such sources as advertisements and textbooks. List them below.

    a.

    b.

    c.

3. Investigate the origin of the metric system. When and where was it first introduced? How are its basic units (the meter, the liter, the gram, the celsius degree) defined?

ACTIVITY 11.1A - MEASURING ANGLES

Materials _required_: _Protractor from the Appendix_

For each sector of the circle graph below,

1. Classify each central angle as acute, obtuse, or right.

2. Measure each angle using a protractor.

3. Record the results in the chart.

SECTOR	TYPE OF ANGLE	MEASUREMENT
PRIMARY		
INTERMEDIATE		
JR. HIGH		
SR. HIGH		

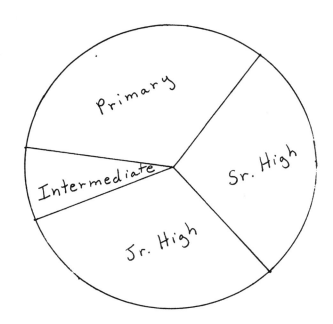

ACTIVITY 11.1B - INTERIOR ANGLES OF POLYGONS

*Materials **required**: Scissors, tracing paper, ruler*

I. A. Trace the following triangle and cut it out. Be certain to number the angles as shown.

   B. Tear off the numbered angles of the triangle and put their vertices together and their sides adjacent.

   C. What kind of angle is formed? What is its measure?

   D. SUMMARY: The sum of the measures of the angles of a triangle is _____.

II. A. Trace the following quadrilaterals. Be certain to number the angles as shown.

   B. Tear off the numbered angles of each quadrilateral and put its vertices together and the sides adjacent.

   C. What kind of angles are formed?

   D. SUMMARY: The sum of the measures of the angles of a quadrilateral is _____.

III. Any polygon can be partitioned into triangles. To form the least number of triangles we select one vertex and connect it with each of the other vertices. The sum of the angles of each of the triangles is 180 ° . Partition each of the following polygons to determine the sum of the vertex angles.

POLYGON	NUMBER OF SIDES	NUMBER OF TRIANGLES	ANGLE SUM

What is the relationship between the number of sides and the number of triangles formed?

Suppose your polygon had n sides. What would be the sum of the vertex angles?

235

ACTIVITY 11.1C - THE MEDIAN

Materials required: Compass, straightedge

I. Define "median" in your own words.

II. To determine the midpoint of AB construct the _ of AB.

III.Construct the three medians of the triangle below.

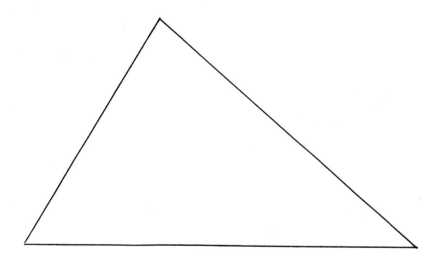

IV. Describe the relationship among the medians.

ACTIVITY 11.1D - PROPERTIES OF A RHOMBUS

Materials required: **Compass, straightedge, protractor**

I. Construct a rhombus.

   1. Construct the perpendicular bisector of line segment AB below.

   2. Label the intersection of the two lines P.

   3. Mark off equal lengths PC and PD along the perpendicular bisector, one above the given line segment and one below the given line segment.

   4. Connect the endpoints of the segments in succession.

   5. The resulting figure is a rhombus.

   6. Use your protractor to check that opposite angles have equal measures and are bisected by the diagonals.

A ●————————————————————————————● B

II. Check the properties of a rhombus.

    1. How do we know that the figure below is a rhombus?

    2. Use your compass and protractor to bisect each of the four interior angles of the rhombus below.

    3. What do you notice about the bisectors of opposite angles?

    4. Measure the angle at the point at which the diagonals intersect. What kind of line segments are these?

    5. Measure the segments into which the point of intersection divides the diagonals. How do they compare?

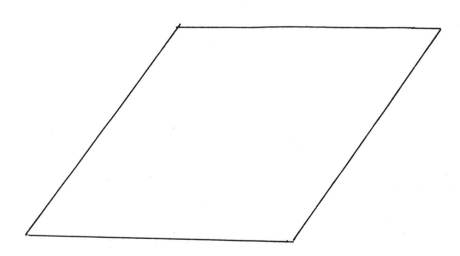

III. SUMMARY: List the four properties of a rhombus.

    1.

    2.

    3.

    4.

ACTIVITY 11.1E - THE GOLDEN RATIO

Materials *required*: *Graph paper, compass, straightedge, ruler*

I. Constructing the golden rectangle

    1. On a sheet of graph paper, draw square ABCD of side 20 units.

    2. Locate midpoint E of side AB.

    3. Draw an arc using E as the center and EC as the radius.

    4. Extend AB to intersect with the arc which you drew.

    5. Label the point of intersection F.

    6. Construct FG perpendicular to AF. Let point G be on the extension of DC.

    7. AFGD is the golden rectangle.

II. Investigating the golden rectangle

    1. Show that $\dfrac{AF}{FG} = \dfrac{(AE + EF)}{FG} = \dfrac{(1 + \sqrt{5})}{2}$

    2. What is the approximate value of $\dfrac{AF}{FG}$? This is the golden ratio.

III. The golden ratio in use.

Measure the length and width of four rectangular objects in your environment. Then divide the length by the width. (Let the length be the longer side.) Record your results in the table below. What do you observe? How can you explain this?

OBJECT	L	W	$\dfrac{L}{W}$
1.			
2.			
3.			
4.			

11.1 CHECK YOUR UNDERSTANDING

1. Measure the following angle and classify it as right, obtuse, or acute.

2. Determine the sum of the interior angles of a regular dodecagon.

3. Determine the measure of each interior angle of a regular polygon with 24 sides.

4. What is true of the medians of any triangle?

5. List four properties of a rhombus.

   a.

   b.

   c.

   d.

6. What is meant by the golden ratio? Name one of its applications.

ACTIVITY 11.2A - PERIMETER

Materials required: *Geoboard (or dot paper)*

Define one unit as the vertical or horizontal distance between two nails on the geoboard.

I. Form the following figures on your geoboard and determine their perimeters.

a.
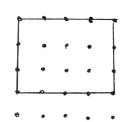

b.
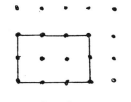

c.
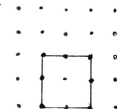

d.

e.
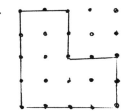

f.
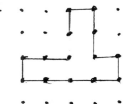

g.
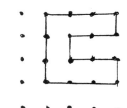

h.
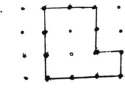

i.
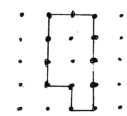

II. Form the following figure on your geoboard. Can you determine the perimeter? Is the distance between two nails diagonally equal to one unit? Explain.

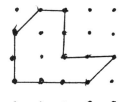

ACTIVITY 11.2B - CIRCUMFERENCE

Materials _required_: *String, ruler, calculator, ten circular objects*

DIRECTIONS: Measure the diameter and circumference of ten circular objects. You may use string to mark off the circumference and then measure the length of the string using a ruler. Divide the circumference by the diameter and round to the nearest hundredth..

OBJECT	DIAMETER(d)	CIRCUMFERENCE(C)	$\frac{C}{d}$
1.			
2.			
3.			
4.			
5.			
6.			
7.			
8.			
9.			
10.			

Using a calculator determine an approximation of π to the nearest hundredth. For every circle the ratio, $\frac{C}{d}$, is equal to π. Compare your calculator approximation with those you derived above. How might the variation be explained? Was the approximation closer for large objects or for small ones?

ACTIVITY 11.2C - AREA AS COVERING

Materials required: *Square tiles from the Appendix, Grids A and B from the Appendix*

I. Determine the area of each of the following figures by covering them with the square tiles.

   1.

   2.

II. A. Some figures are more difficult to cover with square units because of their irregular shapes. We can still determine an approximation of their areas by using the following steps.

    1. Count the number of square units which are completely within the figure. This is the lower bound of the area.

    2. Count the number of square units which are intersected by the figure. Add this number to the lower bound. This is the upper bound of the area.

    3. Now average the two bounds. This is a reasonable approximation of the area.

  B. Now use this technique on the figure on Grid A.

    Lower Bound:

    Upper Bound:

    Average:

  C. Now use this technique on the figure on Grid B.

    Lower Bound:

    Upper Bound:

    Average:

D. Compare the figures which appear on the two grids.

E. Compare the squares on the grids. How many of the squares on Grid B are contained in one of the squares on Grid A?

F. Change the units in Part C by dividing by the number in Part E.
   Lower Bound:
   Upper Bound:
   Average:

G. Compare the approximations in Parts B and F. Which is more accurate? Explain.

H. Suppose we wanted a more accurate approximation. What could we do to determine the area of the given shape?

ACTIVITY 11.2D - AREA FORMULAS

Materials required: *Scissors, area grids from the Appendix*

I. Area of a Rectangle

    A. Count the number of squares in the length of each rectangle, the number of squares in the width of each rectangle, and the total number of squares in the interior of each rectangle. Record this information below.

RECTANGLE #	length(l)	width(w)	area(A)
1			
2			
3			
4			
5			

    B. What is the relationship among the length, the width, and the area of each rectangle? Write the formula below.

        A =

II. Area of a Square

    A. Count the number of squares on a side of each square and the total number of squares in the interior of each square. Record the information below.

SQUARE #	length of side(s)	area(A)
1		
2		
3		
4		
5		

    B. What is the relationship between the length of a side and the area? Write it as a formula below.

        A =

III. Area of a Parallelogram

    A. Cut out the parallelogram.

    B. Cut along the altitude and move the triangular piece to the opposite side of the parallelogram to form a rectangle.

    C. The length of the base (b) of the parallelogram is now the _____ of the rectangle.

    D. The length of the altitude (h) of the parallelogram is now the _____ of the rectangle.

    E. Instead of using A = lw, we will write A =

IV. Area of a Triangle

    A. Cut out the triangles.

    B. Compare the triangles. Are they identical?

    C. Put the triangles together to form a parallelogram.

    D. How could we determine the area of the parallelogram?

    E. What fraction of the parallelogram is the triangle?

    F. Therefore, the area of the triangle(A) =

V. Area of a Trapezoid

    A. Cut out the trapezoid.

    B. Cut along the diagonal.

    C. How could we represent the area of each triangle?

        A_1 =

        A_2 =

    D. Therefore, the area of the trapezoid = A_1 + A_2 = _____ + _____ .

    E. Notice that the height, h, is identical for both triangles. Factor out the GCF for the two expressions. Write the result below.

        A =

VI. Area of a Circle

    A. Cut out the circle.

    B. The circle has been partitioned. Cut apart the sections.

    C. Place the pieces together as indicated below.

    D. The resulting figure approximates a rectangle. The smaller we make the partitions the more it will look like a rectangle.

    E. The radius of the original circle is now the _____ of the "rectangle."

    F. Half of the circumference is at the top of the "rectangle" and half of the circumference is at the bottom of the "rectangle." Therefore, $\frac{1}{2}$ C is the _____ of the "rectangle."

    G. The area of the "rectangle" is l × w, or _____ × _____.

    H. But, C = πd or 2πr. Substitute 2πr for C in the formula in Part G. Simplify. Write the result.   A =

ACTIVITY 11.2E - SQUARE UNITS

**Materials** required: **Masking tape or chalk, yard stick, meter stick**

I. A. Using masking tape(or chalk) and a yard stick, mark off a square with each side equal to one yard.

   B. Label the square, "One square yard."

   C. 1. Each side of the square is _ _ _ _ feet.

   2. The area of the square is _ _ _ _ feet × _ _ _ _ feet, or _ _ _ _ square feet.

   3. Label the square, " _ _ _ _ square feet."

   D. 1. Each side of the square is _ _ _ _ inches.

   2. The area of the square is _ _ _ _ inches × _ _ _ _ inches, or _ _ _ _ square inches.

   3. Label the square, " _ _ _ _ square inches."

   E. Therefore, 1 square yard = _ _ _ _ square feet = _ _ _ _ square inches.

II. A. Using masking tape(or chalk) and the meter stick, mark off a square with each side equal to one meter.

   B. Label the square, "One square meter."

   C. 1. Each side of the square is _ _ _ _ decimeters.

   2. The area of the square is _ _ _ _decimeters × _ _ _ _ decimeters, or _ _ _ _ dm^2 .

   3. Label the square, " _ _ _ _ dm^2 . "

   D. 1. Each side of the square is _ _ _ _ centimeters.

   2. The area of the square is _ _ _ _ centimeters × _ _ _ _ centimeters, or _ _ _ _ _ _ cm^2 .

   3. Label the square " _ _ _ _ _ _ cm^2 ."

   E. Therefore, 1 m^2 = _ _ _ _ _ _ dm^2 = _ _ _ _ _ _ cm^2 .

III. Complete each of the following statements and explain how the equivalence could be demonstrated.

   A. 1 $ft.^2$ = _ _ _ _ $in.^2$

   B. 1 dm^2 = _ _ _ _ _ cm^2

ACTIVITY 11.2F - AREA ON THE GEOBOARD

Materials required: Geoboard, rubber bands

I. Form each of the following figures on your geoboard and determine the area by partitioning the figure into square units using small rubber bands. Let one square unit be defined as the square with a perimeter of four units. (One unit is defined as the distance between two nails vertically or horizontally.)

    a.   b.

II. EXAMPLE: Determine the area of the following figure.

   1. Notice that one side of the figure is neither vertical nor horizontal. Consider it a diagonal of a rectangle. Form the rectangle and determine its area. The area of the piece which is included in the figure is half the area of the rectangle.

   2. Count the remaining square units.

   3. Therefore, the area of the figure is _ _ _ _ _ _ _ square units.

Use this method to determine the area of each of the following figures.

a.   b.

c.   d.

III. The areas of some figures are especially difficult to determine. There is a relationship which will allow us to determine the area of any figure on the geoboard. In order to discover this relationship, form the following figures on your geoboard and determine their areas. Then count the number of Boundary Points (nails or pegs on the perimeter) and the number of Interior Points (nails or pegs in the interior of the figure). Complete the chart below.

FIGURE	AREA(A)	BOUNDARY POINTS(BP)	$\frac{BP}{2}$	INTERIOR POINTS(IP)
A.				
B.				
C.				
D.				
E.				

What is the relationship among A, BP, and IP? This is Pick's Theorem. State the formula below.

A =

251

ACTIVITY 11.2G - SURFACE AREA

Materials __required__: Surface area models from the Appendix

I. DIRECTIONS: 1. Classify the figure as a prism or pyramid.

2. Determine the surface area of each model by counting the number of square units on each face of the model.

3. Use the appropriate area formulas for each face to represent the surface area of the model. Add them together and simplify if possible.

MODEL	CLASSIFY FIGURE	SURFACE AREA	FORMULA
1			
2			
3			

II. DIRECTIONS: 1. Select three objects from your environment.

2. Explain how you could determine the surface area of each of the objects.

OBJECT	SURFACE AREA DETERMINED BY ...
1	
2	
3	

11.2 CHECK YOUR UNDERSTANDING

1. Determine the perimeter of the following figure.

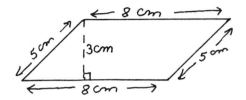

2. What does π represent?

3. What does area represent?

4. Explain how you could derive the formula for the area of a parallelogram.

5. How many square inches are in a square foot?

6. What is Pick's Theorem?

7. In your own words, explain what surface area represents.

ACTIVITY 11.3A - USING LITERS & GRAMS

Materials required: **Items labeled in ml, l, mg, g, or kg**

I. A liter is approximately equal to a quart. Find five items that are labeled with milliliters (ml) or liters (l). List them below with their measurements.

ITEM	MEASUREMENT
1.	
2.	
3.	
4.	
5.	

II. A gram is approximately equal to the weight of a paperclip. Find five items which are labeled with milligrams (mg), grams (g), or kilograms (kg). List them below with their measurements.

ITEM	MEASUREMENT
1.	
2.	
3.	
4.	
5.	

III. Are mass and weight synonymous? Explain.

ACTIVITY 11.3B - RELATIONSHIP AMONG UNITS

Materials required: *Cardboard, tagboard, or heavy paper, plastic bag, tape, one (or two) liter bottle*

Balance scale (optional)

The following is a square decimeter (dm^2) because each of its dimensions is one decimeter.

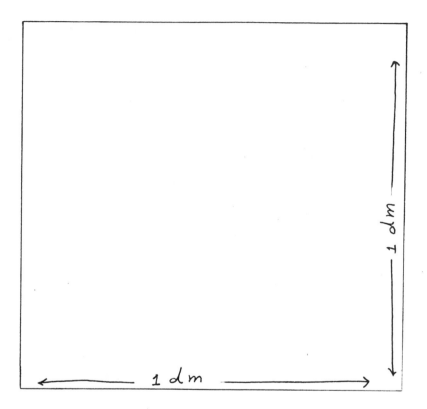

1. Using the dm^2 above as a pattern cut five such squares from cardboard (or heavy paper).

2. Tape the squares together to form a box which is one cubic decimeter (dm^3).

3. Put a plastic bag inside the box.

4. Fill the plastic bag with one liter of water.

5. For water at room temperature 1 dm^3 holds 1 liter and weighs 1 kilogram.

   NOTE: You may verify the weight of the water using a balance scale. Be certain to subtract the weight of the box.

6. If the boxes made by your classmates were added to your box, how many cubic decimeters would you have? How many kilograms? How many liters?

ACTIVITY 11.3C - MEANING OF VOLUME

Materials _required_: _Uncooked navy beans, uncooked kidney beans, a small empty can_

I. Determine the volume of the can using a kidney bean as your basic unit of measure.

   A. Fill the can with kidney beans.

   B. Empty the kidney beans from the can into a bowl and count them.

   C. The volume of the can is _ _ _ _ _ _ _ kidney beans.

II. Determine the volume of the can using a navy bean as your basic unit of measure.

   A. Fill the can with navy beans.

   B. Empty the beans from the can into a bowl and count them.

   C. The volume of the can is _ _ _ _ _ _ _ navy beans.

III. Compare the units.

   A. Did the can hold more navy beans or more kidney beans?

   B. Which measure do you think is more accurate? Why?

IV. Determine the volume of the can using a formula.

   A. Using string and a ruler (or a tape measure) measure the circumference and diameter of the base of the can in centimeters.

         Circumference (C) = _ _ _ _ _ _ _ _ _ _ cm

         Diameter (d) = _ _ _ _ _ _ _ _ _ cm so radius (r) = _ _ _ _ _ _ _ _ cm

     Area of the base (B) of the can is πr^2 **or** _ _ _ _ _ _ _ _ **cm^2**.

   B. Measure the height of the can in centimeters using your metric ruler.

         Height (h) = _ _ _ _ _ _ _ _ _ cm

   C. The volume of the can is found by multiplying the area of the base (B) by the height (h).

         Volume (V) = Bh = _ _ _ _ _ _ _ cm^3

IV. Why is volume normally measured in cubic units?

ACTIVITY 11.3D - VOLUME FORMULAS

Materials required: *Navy beans, cone, cylinder, pyramid, & prism models from the Appendix*

I. Cut out and assemble the four models found in the Appendix.

II. Compare the volumes of the pyramid and the prism.

   A. Fill the pyramid with navy beans.

   B. Pour the beans from the pyramid into the prism.

   C. Repeat steps A and B until the prism is full.

   D. How many times did you pour the beans from the pyramid to the prism?

   E. How does the volume of the pyramid compare to the volume of the prism?

       Volume of the pyramid = _ _ _ _ _ _ volume of the prism.

   F. Compare the heights and bases of the pyramid and the prism.

       Height of the pyramid _ _ _ _ height of the prism

       Area of the base of the pyramid _ _ _ _ area of the base of the prism

   G. Conclusion: If a prism and a pyramid have the same height and their bases have the same area, then Volume of the pyramid = _ _ _ _ _ _ _ Volume of the prism.

   H. Find (or make) a pyramid and determine the volume using the formula.

III. Compare the volumes of the cone and the cylinder.

   A. Fill the cone with navy beans.

   B. Pour the beans from the cone into the cylinder.

   C. Repeat steps A and B until the cylinder is full.

   D. How many times did you pour the beans from the cone to the cylinder?

   E. How does the volume of the cone compare to the volume of the cylinder?

       Volume of the cone = _ _ _ _ _ _ Volume of the cylinder

   F. Compare the heights and bases of the cone and the cylinder.

       Height of the cone _ _ _ _ height of the cylinder

       Area of the base of the cone _ _ _ _ area of the base of the cylinder

   G. Conclusion: If a cone and a cylinder have the same height and their bases have the same area, then volume of the cone = _ _ _ _ _ volume of the cylinder.

   H. Find (or make) a cone and determine the volume using the formula.

ACTIVITY 11.3E - DETERMINING VOLUMES

Materials required: **Objects from the environment, necessary measurement instruments**

1. Select four objects from your environment.

2. Are the objects cones, cylinders, prisms, or pyramids?

3. Name each figure.

4. Determine the volume of each using the necessary measurements and a formula.

FIGURE NUMBER   NAME   VOLUME BY FORMULA

1.

2.

3.

4.

ACTIVITY 11.3F - SPHERES

Materials required: _An orange or a grapefruit, knife, metric ruler, metric caliper from Appendix_

1. Slice the orange or grapefruit in half.

2. The cross section is a great circle. Measure its radius.

3. Determine the area of the interior of the great circle using πr^2.

4. Carefully remove the peel from one of the halves. Divide it in half. This is one fourth of the entire peel.

5. Compare the fourth of the peel to the area of the interior of the great circle. How do they compare? Will the peel cover the interior of the great circle?

6. Since one fourth of the entire peel approximately covers the interior of the great circle, we can write,
$$\pi r^2 = \tfrac{1}{4} S,$$ where S is the surface area of the sphere (i.e., the total amount of peel)

7. Multiplying both sides of the equation in #6 by four will produce the equation,
$$S = 4\pi r^2.$$

8. Imagine that the orange had been divided into sections. The volume of the orange (or grapefruit) would be equal to the sum of the sections. Slice a section into halves, as shown below.

9. Imagine half of a section is a pyramid. The volume would be the sum of all such sections. The volume of each section can be determined by using the formula, $V = \tfrac{1}{3} A_b h$. The height of each section is equal to the radius of the orange (or grapefruit). If we add the A_b of each of the sections together, we will have the entire surface area, S. Substituting, we have,
$$V = \tfrac{1}{3} Sh = \tfrac{1}{3}(4\pi r^2)r = \tfrac{4}{3}\pi r^3.$$

10. Use the metric caliper found in the Appendix to measure the diameter of several other spheres. The radius is half the length of the diameter. Use the formulas and the radius to determine the volume and surface area of each of the spheres. Record the results in the table below.

ITEM	DIAMETER	RADIUS	VOLUME	SURFACE AREA
1.				
2.				
3.				
4.				
5.				
6.				

11. Suppose that you did not have the metric caliper. How could you determine the radius using a piece of string?

ACTIVITY 11.3G - MEASURING TEMPERATURE

I. A. Complete the following chart.

	Fahrenheit Scale	Celsius Scale
Freezing point of water		
Normal human body		
Boiling point of water		

B. How might we use the information in Part A to decide whether to wear a sweater when the temperature is 35° Celsius?

II. A. Temperatures may be converted from Celsius degrees to Fahrenheit degrees by using the formula, $F = \frac{9}{5} C + 32$. Use this formula to change each of the following Celsius temperatures to Fahrenheit degrees.

1.   11°

2.   26°

3.   −8°

4.   42°

5.   57°

B. If we know the temperature in Celsius degrees, we can estimate the equivalent Fahrenheit temperature by using the formula, $F = 2C + 30$. Estimate the Fahrenheit equivalents of each of the temperatures in Part A using this method. Also record the difference between the estimate and your result from Part A. For example, if the estimate is 2 degrees greater, write +2 and if it is 2 degrees less write −2.

	ESTIMATE	COMPARISON WITH PART A RESULT
1.		
2.		
3.		
4.		
5.		

C. Compare the formulas used in Parts A and B. Why does F = 2C + 30 provide a fairly accurate estimate?

III. A. Temperatures may be converted from Fahrenheit degrees to Celsius degrees be using the formula, $C = \frac{5}{9} (F - 32)$. Use the formula to change each of the following Fahrenheit temperatures to Celsius degrees. Round to the nearest tenth.

    1.   30°

    2.   62°

    3.   120°

    4.   95°

    5.   −4°

B. Determine a way in which to estimate the Celsius equivalent of a Fahrenheit temperature. Write it as a formula:

Use your formula to estimate the Celsius equivalent of each Fahrenheit temperature in Part A. Record the difference between the estimate and your result from Part A. For example, if the estimate is two degrees greater, write +2 and if it is 2 degrees less write −2.

ESTIMATE	COMPARISON WITH PART A RESULT
1.	
2.	
3.	
4.	
5.	

11.3 CHECK YOUR UNDERSTANDING

1. Select the most appropriate unit (l, ml, kg, g) for measuring each of the following.

   a. your weight

   b. a few raisins

   c. gasoline in a car's full tank

   d. a glass of milk

   e. a large bag of dogfood

2. Complete the following statement for water at room temperature.

   3 kilograms = _____ liters = _____ dm^3 = _____ cm^3

3. Compare the volumes of a cylinder and a cone with bases of equal size and the same height.

4. Compare the surface area of a sphere with the area of its great circle.

5. Estimate the Fahrenheit equivalent of 38° C.

12: GEOMETRIC CONNECTIONS

Introduction: Many everyday problems require geometric ideas for their solution. These notions are not only practical, but they are very much related to ideas which are not geometric in nature. Through the activities in this section we will see the connection between problem solving, algebra, and technology.

I. We need to rent additional tables for a party.

If we have fifteen guests, then we will rent one table.

If we have fifty-one guests, then we will rent ten tables.

Determine the number of tables (n) which we will need to rent if we have thirty-five guests.

Let p be the number of people we can sit at one table. BE CERTAIN TO ADD YOURSELF TO THE NUMBER OF PEOPLE TO BE SEATED.

    A. Determine the answer algebraically.

    B. Determine the answer geometrically.

II. Research one of the following as it relates to geometry.

Cartography

Geography

Surveying

Navigation

III. Congruency and Similarity

A. Name two objects that have exactly the same size and shape.

B. Name two objects that have the same shape but different size.

ACTIVITY 12.1A - THE DISTANCE FORMULA

*Materials **required**: Graph paper, straightedge*

I. EXAMPLE:

    A. Plot the following points on graph paper.

        1. A(3,2)

        2. B(10,2)

        3. C(10,6)

    B. Draw line segments AB, BC, and AC to form $\triangle$ ABC.

    C. Determine the following lengths by representing them as a difference of two coordinates.

        1. AB = ___ – ___ = ___

        2. BC = ___ – ___ = ___

    D. Use the Pythagorean Theorem to determine AC.

II. The General Form

    A. Consider the following diagram.

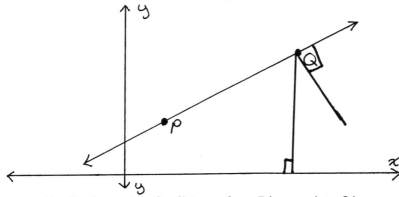

    B. To determine the distance from $P(x_1 , y_1)$ to $Q(x_2 , y_2)$:

        1. Drop a perpendicular from Q to form $\triangle$ PQ.

2. What are the coordinates of point R?

3. Represent the length of QR as the difference of two coordinates.

4. Represent the length of PR as the difference of two coordinates.

5. Use the Pythagorean Theorem to represent the length of side PQ.

6. Therefore, the length of side PQ represents the distance from P to Q.

   $PQ = d =$

III. Application:

Determine the perimeter of the triangle with vertices at A(4, 8), B(9, 10), and C(− 2, 3).

1. Graph the points.

2. Connect the points to form $\triangle$ ABC.

3. Estimate the perimeter.

4. Use the distance formula as necessary to determine the lengths of the sides.
   (Use a calculator to estimate any irrational numbers to the nearest tenth.)

   AB =

   AC =

   BC =

5. The perimeter of the triangle is _____.

ACTIVITY 12.1B - SLOPE

Materials _required_: _Graph paper_

I. Graph the following systems of equations.

A. $3x + 2 = y$

       $3x - 10 = y$

    1. What kind of lines are these?

    2. What is the slope of each line?

    3. Conclusion:

B. $4x - 5 = y$

       $-\frac{1}{4}x + 2 = y$

    1. What kind of lines are these?

    2. What is the slope of each line?

    3. Conclusion:

II. A. Plot the following points:

    A: (1, 3)   B: (4, 11)   C: (12, 8)   D: (9, 0)

B. What figure is formed when you connect the points in alphabetical order in Part A?

C. Draw the diagonals. Determine the slope of each diagonal. What is the relationship? What kind of angles do the diagonals form?

D. Conclusion:

III. Plot the following points:

    (1, 4)   (3, 8)   (9, 5)   (7, 1)

A. What figure is formed?

B. Find the slope of each side?

C. How could we use the slope to show that the points above are NOT are the vertices of a trapezoid?

D. How could we use the slope to show that the points above represent a rectangle and not just a parallelogram?

ACTIVITY 12.1C - EQUATIONS OF CURVES

Materials _required_: **Graph paper**

I. Plot each of the following equations on the same set of axes.

    A. $y = x^2$

    B. $y = (x + 3)^2$

    C. $y = (x - 2)^2$

    D. $y = (x - 4)^2$

    What is the relationship among these equations?

    Predict the graph of $y = (x + 7)^2$.

II. Plot each of the following equations on the same set of axes.

    A. $y = x^2 + 3$

    B. $y = x^2 - 2$

    C. $y = x^2 + 4$

    What is the relationship among these equations?

    Predict the graph of $y = x^2 - 6$.

III. Plot each of the following equations on the same set of axes.

    A. $y = (x + 3)^2 + 2$

    B. $y = (x - 2)^2 + 4$

    C. $y = (x - 4)^2 + 5$

    What is the relationship among these equations?

    Predict the graph of $y = (x + 1)^2 - 8$.

IV. SUMMARY: In general, describe the graph of $y = (x - h)^2 + k$.

ACTIVITY 12.1D - EQUATION OF A CIRCLE

Materials required: *Graph paper*

By definition, a circle is a set of points equidistant from a fixed point called the center. The equation of a circle will, therefore, be determined by the distance from the center to any point on the circle.

I. The equation

    A. Consider the following diagram.

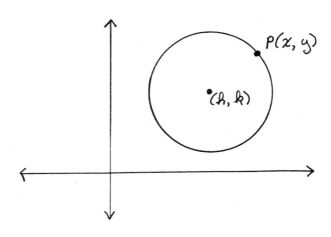

    B. What are the coordinates of the center of the circle?

    C. What are the coordinates of point P, on the circle?

    D. Determine the distance from P to the center using the distance formula. Call this distance r. It is radius of the circle.

    E. Square both sides of the equation to obtain an expression equal to r^2.

274

II. Application

    A. Given the equation, $(x - 2)^2 + (x + 4)^2 = 25$.

        1. What is the center of the circle?

        2. What is the length of the radius?

        3. Sketch the circle.

        4. Name two points on the circle.

        5. Using the distance formula, determine the distance from each of the two points which you named in #4 to the center of the circle. Compare each distance to the length of the radius.

    B. Draw a line tangent to the circle at the point (5,0).

        1. Determine the slope of the radius drawn to this point.

        2. Determine the slope of the tangent at this point.

        3. Compare the slopes in #1 and #2. What kind of lines are these?

        4. Write the equation of the tangent using the slope and the coordinates of the point at which the line is tangent.

12.1 CHECK YOUR UNDERSTANDING

1. Determine the distance from (3, 8) and (14, 15).

2. a. If two lines are parallel, then their slopes are _____.

   b. If two lines are perpendicular, then their slopes are _____.

3. Describe the graph of $y = (x + 8)^2 - 4$.

4. Describe the graph of $(x - 10)^2 + (y + 6)^2 = 100$.

ACTIVITY 12.2A - SIMILAR TRIANGLES

Materials **required:** *Compass, straightedge, protractor, ruler*

Two triangles are similar if corresponding angles are equal.

I. Investigation #1.

1. Using your compass and straightedge copy ∠ A at point D on line segment DE.

2. Using your compass and straightedge copy ∠ B at point E on line segment DE.

3. Extend the sides of ∠ D and ∠ E until they intersect. Label the point of intersection F.

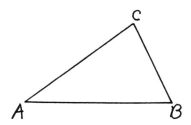

4. Using your protractor measure ∠ C and ∠ F. How do they compare?

5. Using your ruler measure the sides of each triangle. Record their lengths below.

   a. AB = _ _ _ _ cm   DE = _ _ _ _ cm $\dfrac{AB}{DE}$ = _ _ _ _ cm

   b. AC = _ _ _ _ cm   DF = _ _ _ _ cm $\dfrac{AC}{DF}$ = _ _ _ _ cm

   c. BC = _ _ _ _ cm   EF = _ _ _ _ cm $\dfrac{BC}{EF}$ = _ _ _ _ cm

6. Compare the ratios in #5. Are they equal?

7. Are the triangles similar? Explain.

8. SUMMARY: If _ _ _ _ _ _ _ _ _ _ _ _ _ of one triangle are equal to _ _ _ _ _ _ _ _ _ _ _ _ of another triangle, then the triangles _ _ _ _ _ _ _ _ _ _ _ _ _ _ _ .

II. Investigation #2.

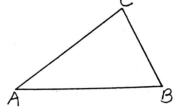

1. Measure the sides of △ ABC to the nearest tenth of a centimeter.

    a. AB = _ _ _ _

    b. BC = _ _ _ _

    c. AC = _ _ _ _

2. Multiply each of the lengths in #1 by 3. Name these lengths in the following way.

    a. 3AB = _ _ _ _ = DE

    b. 3BC = _ _ _ _ = EF

    c. 3AC = _ _ _ _ = DF

3. Draw line segments DE, EF, and DF with the lengths indicated above.

4. Using your compass and straightedge construct a triangle with sides DE, EF, and DF.

5. Measure the angles of each triangle.

    a. m∠ A = _ _ _ _

    b. m∠ B = _ _ _ _

    c. m∠ C = _ _ _ _

    d. m∠ D = _ _ _ _

    e. m∠ E = _ _ _ _

    f. m∠ F = _ _ _ _

6. What observation can be made about the angles of the two triangles?

7. Are the triangles similar? Explain.

8. SUMMARY: Two triangles are similar if _ _ _ _ _ _ _ _ _ _ _ _ of one triangle are

_ _ _ _ _ _ _ _ _ _ _ _ to _ _ _ _ _ _ _ _ _ _ _ _ of the other triangle.

III. Investigation #3.

    1. Given line segments AB and AC and ∠ A. Using your compass and straightedge construct

       a triangle with ∠ A included between sides with lengths AB and AC.

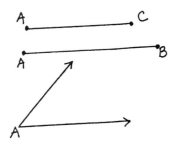

    2. Draw BC to form △ ABC.

    3. Construct line segment DE which equals twice AB and line segment DF which equals twice

       AC.

    4. Construct △ DEF with ∠ D equal to the measure of ∠ A and included between DE and

       DF.

    5. Make the following measurements.

         a. m∠ E = _ _ _ _

         b. m∠ F = _ _ _ _

         c. m∠ B = _ _ _ _

         d. m∠C = _ _ _ _

         e. EF = _ _ _ _

         f. BC = _ _ _ _

    6. Are the triangles similar? Explain.

    7. SUMMARY: Two triangles are similar if two sides of one triangle are _ _ _ _ _ _ _ _ _ _ _ _

       to two sides of another triangle and the angles included between those sides are _ _ _ _ _ _ _.

281

Materials *required*: *Compass, straightedge, ruler, protractor*

Similar polygons have corresponding angles equal and corresponding sides in proportion.

Given: Polygon ABCDEFG

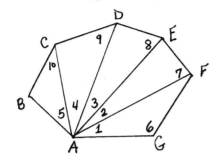

STEP I:

    1. Copy ∠ 1 onto point P on PQ.

    2. Copy ∠ 6 onto point Q on PQ.

    3. Draw △ PQR.

STEP II:

    1. Copy ∠ 2 onto point P on PR.

    2. Copy ∠ 7 onto point R on PR.

    3. Draw △ PSR.

STEP III:

    1. Copy ∠ 3 onto point P on PS.

    2. Copy ∠ 8 onto point S on PS.

    3. Draw △ PTS.

STEP IV:

    1. Copy ∠ 4 onto point P on PT.

    2. Copy ∠ 9 onto point T on PT.

    3. Draw △ PTU.

STEP V:

1. Copy ∠ 5 onto point P on PU.

2. Copy ∠ 10 onto point P on PU.

3. Draw △ PUV.

CHECK:

1. Measure each side of the two polygons.

2. Are the sides in proportion? Explain.

3. Are the two polygons similar? Explain.

ACTIVITY 12.2C - CONGRUENT TRIANGLES

Materials <u>required</u>: *Compass, protractor, straightedge, ruler*

Congruent triangles are triangles whose corresponding angles are equal and whose corresponding sides are equal.

I. Investigation #1.

    1. Using a compass and straightedge construct $\triangle$ ABC with sides equal to the three line segments below.

    2. Using compass and straightedge construct a second triangle, $\triangle$ DEF, with sides equal to the three line segments above.

    3. Complete each statement.

        a. AB =   d. DE =

        b. BC =   e. EF =

        c. AC =   f. DF =

4. Using your protractor measure each of the angles in △ ABC and △ DEF. Record their measures below.

       a. ∠ A =

       b. ∠ B =

       c. ∠ C =

       d. ∠ D =

       e. ∠ E =

       f. ∠ F =

5. What observations can you make about the angles of △ ABC and △ DEF?

6. Are △ ABC and △ DEF congruent? Explain.

7. SUMMARY: Two triangles are congruent if three sides of one triangle are _ _ _ _ _ _ _ _ _ _ to three sides of another triangle.

II. Investigation #2.

1. Copy $\angle$ A onto line l at D.

2. On the sides of $\angle$ D mark off lengths equal to AB and AC. Label the endpoints E and F, respectively.

3. Draw EF.

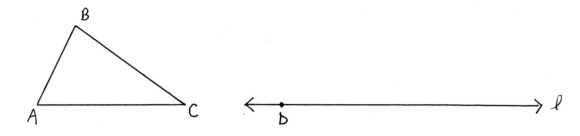

4. Using your protractor measure the angles of each triangle.

    a. $\angle$ B =

    b. $\angle$ C =

    c. $\angle$ E =

    d. $\angle$ F =

5. What observations can you make about the angles of $\triangle$ ABC and $\triangle$ DEF?

6. Using your ruler measure EF and BC to the nearest tenth of a centimeter..

    a. EF =

    b. BC =

7. Are $\triangle$ ABC and $\triangle$ DEF congruent? Explain.

8. SUMMARY: Two triangles are congruent if _____ to

_____ of another triangle.

286

III. Investigation #3.

1. Using your compass and straightedge, copy line segment AB onto line l at D.

2. Label the other endpoint E.

3. Using your compass and straightedge copy ∠ A onto DE at D and ∠ B onto DE at E.

4. Extend the sides of ∠ D and ∠ E and label their point of intersection E.

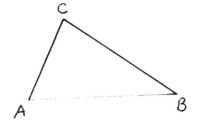

5. Use your protractor to measure ∠ F and ∠ C.

      a. ∠ F =

      b. ∠ C =

6. Using your ruler make the following measurements.

      a. AC =

      b. BC =

      c. DF =

      d. EF =

7. Are △ ABC and △ DEF congruent? Explain.

8. SUMMARY: If _____ of one triangle

    are equal to _____ of another triangle then

    the triangles are congruent.

287

ACTIVITY 12.2D - INDIRECT MEASUREMENT

Materials required: *Mirror, meter stick or metric tape measure.*

1. Record all measurements in the diagram below.

2. Place a mirror on the ground a distance of ten meters from the base of a tree.

3. Move away from the mirror until you can see the highest part of the tree reflected in the mirror.

4. Mark the location on the ground.

5. Measure the distance from that point to the mirror.

6. Measure the height of your own eye level.

7. Using a proportion determine the height of the tree.

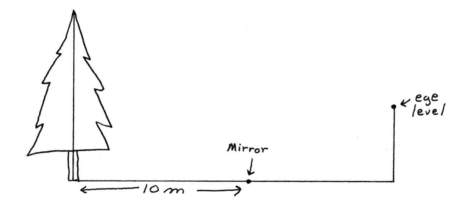

12.2 CHECK YOUR UNDERSTANDING

1. List three ways in which to show that triangles are similar.

    a.

    b.

    c.

2. List three ways in which to show that triangles are congruent.

    a.

    b.

    c.

3. Determine if the following pair of polygons are similar. Explain.

    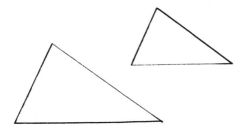

4. Name a pair of properties of similarity and congruency which share the same name. How do they differ in meaning?

ACTIVITY 12.3A-THE REPEAT COMMAND

Materials required: **Computer, LOGO software**

EXAMPLE: Construct a square using LOGO.

Method One: Use FD and RT.

FD 40

RT 90

FD 40

RT 90

FD 40

RT 90

FD 40

RT 90

Method Two: Use the REPEAT command.

REPEAT 4 [FD 40 RT 90]

A. Construct each of the following polygons using each of the methods above.

1. An equilateral triangle with each side equal to 35.

Method One:

Method Two:

2. A regular pentagon with each side equal to 30.

Method One:

Method Two:

291

3. A regular hexagon with each side equal to 25.

    Method One:

    Method Two:

4. A regular octagon with each side equal to 25.

    Method One:

    Method Two:

B. What information did you need in order to draw these polygons?

C. Write a general formula for constructing a polygon of n sides of x units each using the REPEAT command.

D. What happens to the picture as n approaches 360 and x approaches 0?

292

E. Form a design using three different pattern blocks. Draw the design using LOGO. Record your design and the commands below.

F. Form a design using any four pattern blocks. Draw the design using LOGO. Record your design and the commands below.

G. Form a design using any five pattern blocks. Draw the design using LOGO. Record your design and the commands below.

ACTIVITY 12.3B-TESSELLATIONS USING PROCEDURES

Materials required: *Computer, LOGO software*

EXAMPLE: PRESS "OPEN APPLE" - F
         TYPE:
            TO SQUARE
            REPEAT 4 [FD 50 RT 90]
            END
         PRESS "OPEN APPLE" - F

         TYPE:
            PU
            LT 90
            FD 100
            RT 90
            PD
            REPEAT 4 [SQUARE RT 90 FD 50 LT 90]
            PU
            LT 90
            FD 175
            RT 90
            BK 50
            PD
            REPEAT 3 [SQUARE RT 90 FD 50 LT 90]

What is the result? Sketch it below.

PROBLEMS:

1. Create the following polygons with each side equal to 30 as PROCEDURES.

    A. HEXAGON

    B. EQUILATERAL TRIANGLE

    C. OCTAGON

2. Create tessellations such as the following:

    A.

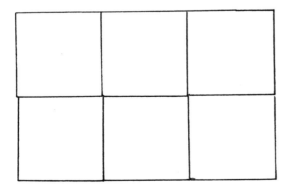

B.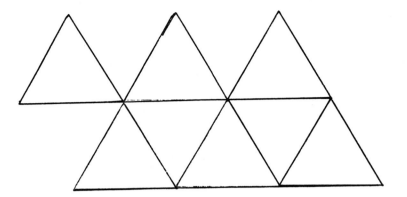

C.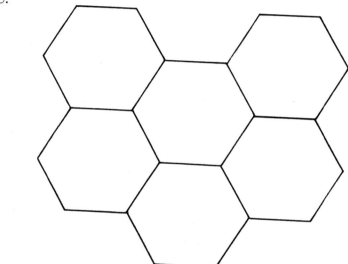

ACTIVITY 12.3C - VARIABLES

Materials required: *Computer, LOGO software*

I. THE SQUARE

    A. Enter the following PROCEDURE.

             TO SQUARE   :SIDE

             REPEAT 4 [FD   :SIDE   RT 90]

             END

    B. TYPE: SQUARE 40

       Describe the result.

    C. If we double the side, will the area double? Predict and check.

    D. If we divide the side by two, will the area be half of the present area? Predict and check.

    E. What will happen to the area if we square the length of a side? Predict and check.

    F. In order to multiply the area by nine, what should we do to the length of a side?

II. THE RECTANGLE

A. Enter the following PROCEDURE:

    TO RECTANGLE   :HEIGHT   :WIDTH
    REPEAT 2 [FD   :HEIGHT RT 90 FD   :WIDTH RT 90]
    END

B. TYPE: RECTANGLE 50 100.

  Describe the result.

C. What happens to the area if we double both dimensions? Predict and check.

D. What happens to the area if we divide the width by two? Predict and check.

E. What will be the result of RECTANGLE 50 0? Predict and check.

ACTIVITY 12.3D - RECURSION

Materials required: **Computer, LOGO software**

I. DEFINE THE PROCEDURE:

     TO SQUARE

     REPEAT 4 [FD 40 RT 90]

     END

Describe the result.

II. Instructions: For each of the following, type the procedure, predict the result, and then run the procedure and describe the result.

    A. TO BOX   PREDICT:

       RT 90

       SQUARE   DESCRIBE:

       END

    B. TO STACK   PREDICT:

       SQUARE

       RT 90

       SQUARE   DESCRIBE:

       END

    C. TO TURN   PREDICT:

       SQUARE

       RT 45

       SQUARE   DESCRIBE:

       END

D. TO CLIMB   PREDICT:
   SQUARE
   RT 180
   SQUARE   DESCRIBE:
   END

III. Use a procedure with recursion to draw each of the following:

A.

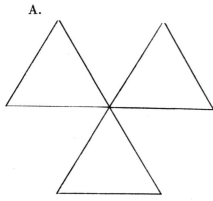

B.

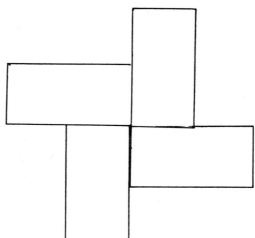

C.

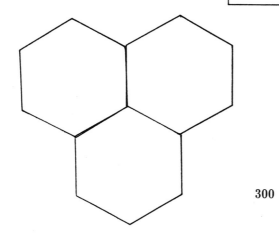

300

12.3 CHECK YOUR UNDERSTANDING

1. Construct a regular octagon of side 30 using the REPEAT command.

2. Create a regular hexagon with side equal to 25 using a PROCEDURE.

3. Explain how you could use a PROCEDURE containing variables to show that doubling the dimensions of a rectangle does not double its area.

4. Explain one way in which to use RECURSION to construct a circle.

13: OTHER GEOMETRIES

Introduction: The activities in this section will illustrate the ways in which Transformational Geometry, Projective Geometry, and Topology are applied to patterns and perspective in our world. An awareness of some basic notions in each area will enrich our understanding and appreciation of many facets of our lives.

I. PATTERNS: Find at least three patterns used in wallpaper, tiling, fabric, or music. Attach them (or reproductions) in the spaces below.

A. Item: _____

B. Item: _____

C. Item: _____

II. PERSPECTIVE: Visit an art museum or a library and consider at least one painting from each of the following periods. Then answer the questions below.

A. The Middle Ages

1. What is the title of the painting?

2. Who was the artist?

3. In approximately what year was it completed?

4. Does it have a central character or object? If so, who or what is it?

5. Does it give the impression of depth, or three dimensions? If so, how?

B. The Renaissance

1. What is the title of the painting?

2. Who was the artist?

3. In approximately what year was it completed?

4. Does it have a central character or object? If so, what is it?

5. Does it give the impression of depth, or three dimensions? If so, how?

ACTIVITY 13.1A - TRANSLATIONS

Materials Required: _Tracing paper, ruler_

1. Trace $\triangle ABC$ and ray XY on a sheet of tracing paper.

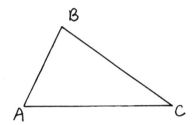

2. Slide the tracing paper in the direction defined by ray XY until the end of the ray is on the "tip" of ray XY.

3. With a sharp pencil, mark the new locations of the vertices A, B, and C. Label them A', B', and C' respectively.

4. $\triangle A'B'C'$ is the image of $\triangle ABC$ under the translation defined by ray AB. In elementary classrooms, a translation is often called a "slide." It may be represented as $\mathbf{S}_{\mathbf{XY}}$.

5. Measure the distances from A to A', from B to B', and from C to C'. What can you conclude?

6. How does $\triangle ABC$ compare to $\triangle A'B'C'$?

7. Draw any pentagon ABCDE on graph paper. If $\mathbf{S}_{\mathbf{PQ}}$ is defined as the translation which moves each point three units horizontally, how could you locate the image of pentagon ABCDE, pentagon $A'B'C'D'E'$, without using tracing paper? Demonstrate this.

8. Check your work in #7 with the technique described in #1-#3.

ACTIVITY 13.1B - REFLECTIONS

Materials <u>Required</u>: Tracing paper, ruler, compass

1. Trace △ABC, line ℓ, and point P on line ℓ.

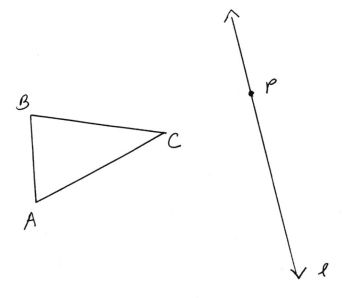

2. "Flip" the tracing paper over. Align the images of line ℓ and point P with the originals.

3. With a sharp pencil, mark the new locations of the vertices A, B, and C. Label them A', B', and C', respectively.

4. △$A'B'C'$ is the image of △ABC under the reflection defined by line ℓ. In elementary classrooms, a reflection is often called a "flip." It may be represented as **M**$_\ell$.

5. Measure the distances from A to A', from B to B', and from C to C'. Is there a relationship between these distances and line ℓ?

6. How does △ABC compare to △$A'B'C'$?

7. Draw any quadrilateral ABCD and any line ℓ on a sheet of unlined paper. Using a compass and straightedge construct perpendiculars to line ℓ through points A, B, C, and D. Extend the perpendiculars through line ℓ. How can you locate the image of each vertex using the relationship described in #5? Demonstrate this.

8. Check your work in #7 with the technique described in #1-#3.

ACTIVITY 13.1C - ROTATIONS

Materials Required: Tracing paper, protractor, straightedge, compass

1. Draw the line segment from point Ơ to vertex A.

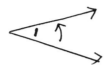

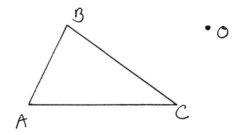

2. Trace the directed angle, ∠ 1.

3. Place the image of the vertex of ∠ 1 on point Ơ.

4. Align the terminal side of ∠ 1 with line segment ƠA.

5. Trace line segment ƠA, △ ABC, and point Ơ.

6. Rotate the image of line segment ƠA until the image of the initial side of ∠ 1 aligns with line
   segment ƠA. (Be certain that you keep point Ơ fixed.)

7. Using a sharp pencil, mark the new locations of the vertices A, B, and C. Label them A', B',
   and C', respectively.

8. Draw the line segment A'Ơ. Measure ∠ AƠA'. How does it compare to the measure of ∠ 1?

9. Draw line segments BƠ, B'Ơ, CƠ, and C'Ơ. Measure ∠ BƠB' and ∠ CƠC'. How do their
   measures compare to that of ∠ 1?

10. Your rotation "turned" the image of $\triangle ABC$ the number of degrees and direction defined by $\angle 1$. The rotation may be represented by $\mathbf{R}_{(\theta, 40)}$.

11. Draw any quadrilateral ABCD and any point P on a sheet of unlined paper. If $\mathbf{R}_{(P, -60)}$ is defined as the rotation which turns each vertex of quadrilateral ABCD 60° clockwise about point P, how could you locate the image of quadrilateral ABCD, quadrilateral $A'B'C'D'$, without using graph paper?

    a. How could you use a protractor and straightedge to find the image? Describe and demonstrate your method.

    b. How could you use a compass and straightedge to find the image? Describe and demonstrate your method.

ACTIVITY 13.1D - GLIDE REFLECTIONS

Materials Required: *Tracing paper*

1. Given below are △ABC, line ℓ, and ray XY.

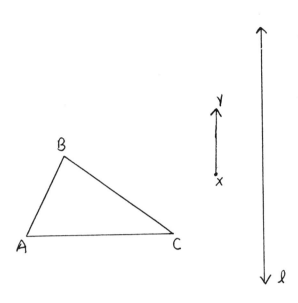

2. Translate △ABC using ray XY and then reflect that image through line ℓ. This glide reflection may be represented by $M_\ell(S_{XY})$. Be certain to label the image △$A'B'C'$.

3. Translate △ABC by reflecting through line ℓ first and then translating that image using XY. This glide reflection may be represented by $S_{XY}(M_\ell)$. Did this image coincide with that found in #2? Will this always be true?

4. Given △ DEF, line , and ray PQ, determine the image which results from each of the following glide reflections.

a. **M(S_{PQ})**

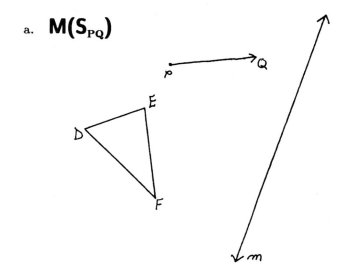

b. **S_{PQ}(M)**

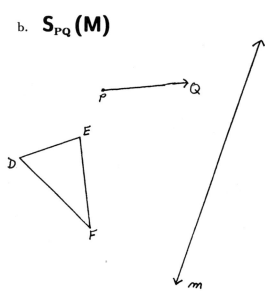

5. Are the results of #4a and #4b identical? Why do you suppose this is true? What would you conclude?

ACTIVITY 13.1E - LINE SYMMETRY

A figure has line symmetry if there exists a line of reflection. This line of reflection divides the figure into two parts which are mirror images of one another. Line symmetry may be verified by folding on the line of reflection.

1. Determine whether each of the following figures has line symmetry.

2. If so, determine the number of lines of symmetry.

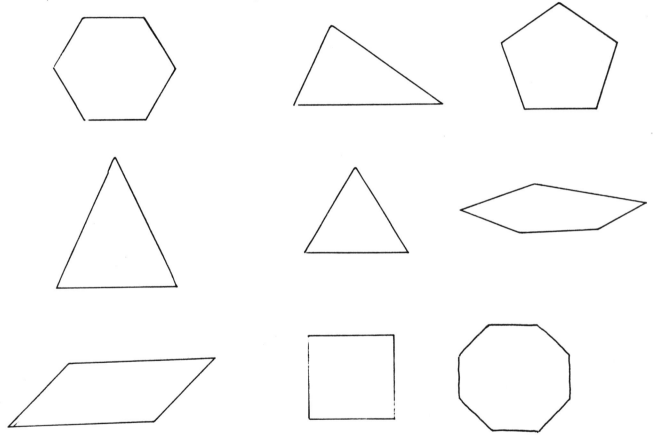

3. Discuss the way in which line symmetry relates to the transformation called reflection.

ACTIVITY 13.1F - POINT SYMMETRY

A figure has point symmetry if it can be rotated about a point in such a way that the image matches the pre-image. The order of the rotational symmetry is the number of turns which are necessary to bring the figure back to its original position.

1. Determine whether each of the following figures has rotational symmetry.

2. If so, determine its order.

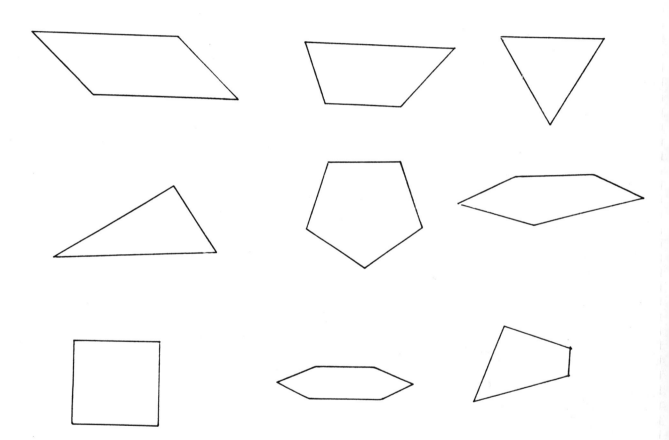

3. Discuss the way in which rotational symmetry relates to the transformation called rotation.

ACTIVITY 13.1G - TESSELLATIONS

Materials Required: Tracing paper

Figures which will completely cover a closed figure without "gaps" or "overlapping" are said to tessellate.

  I. Determine whether each of the following figures will tessellate by tracing it repeatedly.

  II. For each of those figures that tessellates, pick a point at which vertices meet.

    A. Describe the point using the number and kinds of polygons that meet there.

    B. What can be said concerning the sum of the angles which share that vertex point?

1. Isosceles triangle

2. Equilateral triangle

3. Scalene triangle

4. Square

5. Rectangle

6. Rhombus

7. Parallelogram

8. General quadrilateral

9. Pentagon

10. Hexagon

11. Octagon

ACTIVITY 13.1H - CREATE A TESSELLATION

I. Tessellation by Translation:

    A. Change one side of each polygon on the left.

    B. Translate this change to the opposite side of the polygon.

    C. Repeat this change using the grid on the right to show that the result tessellates.

EXAMPLE:

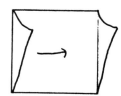
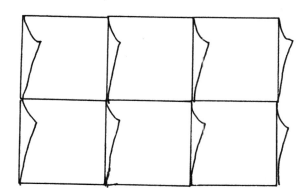

1.

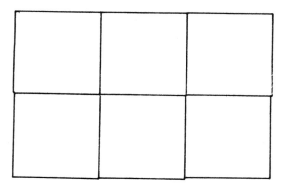

2.

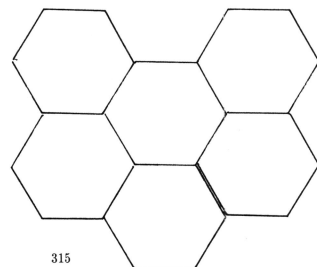

315

II. Tessellation by Rotation

A. Change the side AB of the polygon at the left.

B. Rotate the new side AB at A until it coincides with side AC.

C. Repeat this pattern on the grid at the right to demonstrate that the new figure tessellates.

EXAMPLE:

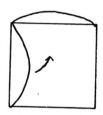

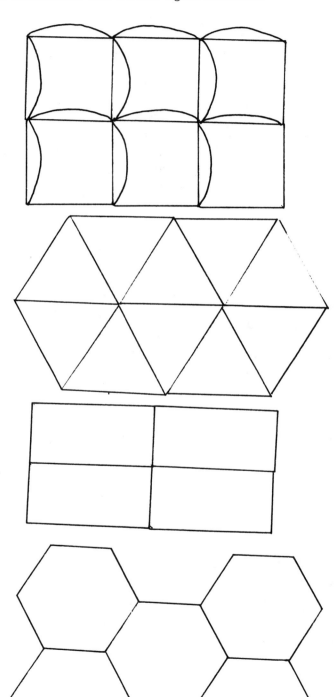

1.

2.

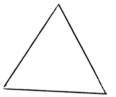

3.

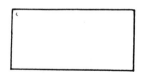

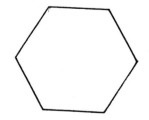

316

13.1 CHECK YOUR UNDERSTANDING

1. Perform each of the following transformations.

a. **R**$_{(P,50)}$

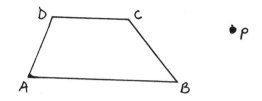

b. **M (S$_{XY}$)**

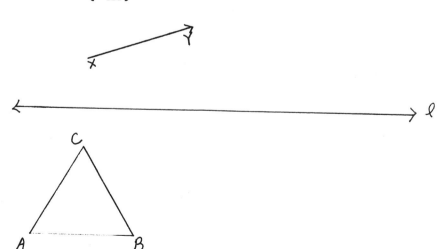

2. What is the relationship between each point and its image under a translation?

3. What is the relationship among each point, its image, and the line of reflection?

4. What is the relationship among each point, its image, and the angle of rotation?

5. Does a nonagon have line symmetry? Explain.

6. Does a dodecagon have rotational symmetry? Explain.

7. Will a heptagon tessellate? Explain.

ACTIVITY 13.2A - PROJECTIVE GEOMETRY

Materials Required: *Ruler*

I. Construct a perspective drawing of the figure above using the distance from X to Y as the distance between the primary and the secondary vanishing points.

II. Construct a perspective drawing of the given figure using four times the distance from X to Y as the distance between the primary and the secondary vanishing points.

III. What effect did the change in distance have on the perspective drawing? Explain.

ACTIVITY 13.2B - TOPOLOGY

Materials Required: Tracing paper, scissors

I. Instructions

    A. Trace each of the following figures.

    B. Cut out each figure and remove the "center" portions.

    C. Complete the table below by investigating the number of cuts which will leave the figure intact.

FIGURE   # OF HOLES   # OF CUTS POSSIBLE

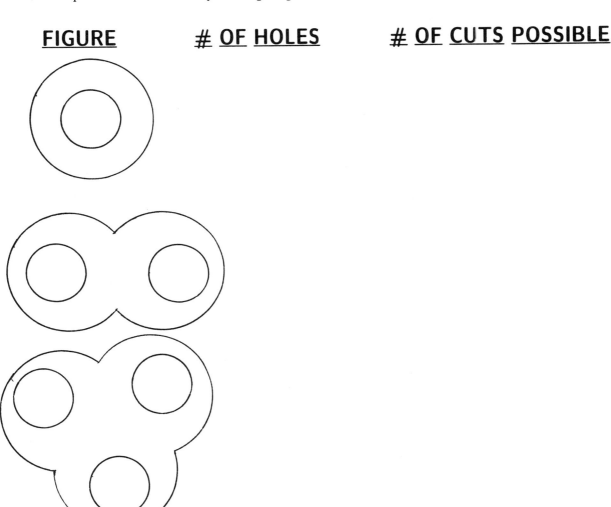

II. Name one item which has the same number of holes as each figure in the chart above. List these topologically equivalent objects below.

321

ACTIVITY 13.2C - NETWORKS

*Materials **required**: Platonic solid models from Chapter 10.2A.*

I. Complete the following chart using your models of the Platonic solids from Chapter 10.

FIGURE	**# OF ODD VERTICES**	**# OF EVEN VERTICES**	**TRAVERSABLE?**
TETRAHEDRON			
HEXAHEDRON			
OCTAHEDRON			
DODECAHEDRON			
ICOSAHEDRON			

II. Is the following floorplan traversable? Explain.

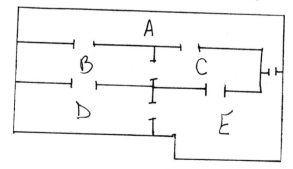

13.2 CHECK YOUR UNDERSTANDING

1. Make a perspective drawing of the following figure using the distance from A to B as the distance between the primary and secondary vanishing points.

2. What effect does the distance between the primary and secondary vanishing points have on our perspective?

3. Name two objects that are topologically equivalent and explain this equivalence.

4. Give an example of a network that is not traversable. Explain.

5. For each of the patterns that you selected at the beginning of this unit, discuss the use of transformations and tessellations.

    A.

    B.

    C.

6. Discuss the use of perspective in the Renaissance painting that you selected at the start of this unit.

APPENDIX 1

ANSWERS TO ACTIVITIES

ANSWERS TO ACTIVITIES

1.1A   **Pages 3 & 4**

1. Answers will vary.

2. a. 6   b. 11

3. a. Triangles: 7;   Quadrilaterals: 2;   Pentagons: 1;

    b. Triangles: At least 6 pairs;   Quadrilaterals: At least 6 pairs;

       Pentagons: At least 4 pairs;   Hexagons: At least 2 pairs;

1.1B   **Pages 5 & 6**

1. 1: V, R;   2: D, R;   3: H, V, D, R;   4: D, R;   5: V;   6: H, V, D, R;

2. Answers will vary.

1.2A   **Pages 9 - 12**

1. a. 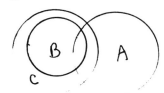   b.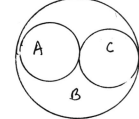

   c. 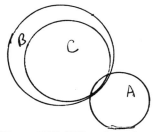   d.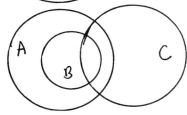

2. a. {YT, yt, YC, YS, ys}   b. { }   c. {YT, RT}   d. {RS, BS, YS}

3. a. {bt, rt}   b. {BT, bt, BS, bs}   c. {BC, bc, BS, bs, bt}   d. { }

4. No, numbers 2 and 3 do not yield the same results.

5. a. {rc, yc, bc, ys, YS, YC, YT, yt}   b. {BC, bc, BT, bt}   c. {RT, YT, BT, bt}

   d. {RS, rs, YS, ys, BS, bs}

6. a. {yt, yc, rc, bc}   b. {BT, bt}   c. {BT, bt}   d. {rs, RS}

7. No, 5a through 5d would have given the same results as 6a through 6d, respectively.

8. a. {yc}   b. { }   c. {BT}   d. {rs}

9. a. The small yellow blocks.   b. The blue blocks which are <u>both</u> circles and triangles.

   c. The large blue blocks.   d. The small red squares.

10. Yes, both expressions imply that all three restrictions hold.

11. a. {YT, YS, YC, yt, ys, yc, rc, bc, bt, bs, rt, rs}   b. {BS, bs, BC, bc, BT, bt}

c. {RT, YT, BT, bt, BC, bc, BS, bs}   d. {YS, BS, ys, bs, rs, RS, RT, rt, rc, RC}

12. a. The set of blocks which are yellow or small.   b. The set of blue blocks.

12. c. The set of blocks which are large triangles or blue.   d. The set of blocks which are squares or red.

13. Yes, you are combining sets together. Set union is associative.

1.2B   Page 13 and 14

1. a. {rt}   b. {yt, bt, RT, BT, YT}   c. {BT, YT}   d. {bt, yt, rt, RT}   e. {BT, YT}
   f. {RT, BT, YT, bt, yt}

2. c and e; b and f;

3. a. { }   b. {YT, yt, YS, ys, YC, yc}   c. {YT, yt, yc}   d. {YC, YS, ys}   e. {YT, yt, yc}
   f. {YT, yt, YS, ys, YC, yc}

4. b and f; c and e:

5. $\overline{A \cap B} = \overline{A} \cup \overline{B}$; $\overline{A} \cap \overline{B} = \overline{A \cup B}$

1.2C   Pages 15 and 16

ANSWERS WILL VARY. EXAMPLES FOLLOW.

1. A = {large blocks}   B = {large red blocks}   C = {yellow blocks}

2. A = {all yellow blocks}   B = {all squares}   C = {all large blocks}

3. A = {red blocks}   B = {blue blocks}   C = {yellow blocks}

4. A = {all triangles}   B = {all squares}   C = {all yellow blocks}

5. A = {all large blue blocks}   B = {all large triangles}   C = {all small blocks}

6. A = {all large blocks}   B = {large blue polygons}   C = {all blue blocks}

7. A = {Blue blocks}   B = {small blue blocks}   C = {small blue polygons}

1.3A   Pages 19 and 20

I. 1. T T   2. T F   3. F T   4. F F

II. A.   ∼ p: The block is not blue.   ∼ q: The block is not a triangle.

     1. T T F F   2. T F F T   3. F T T F   4. F F T T

     ∼ p is false when p is true.   ∼ q is false when q is true.

B. p ∧ q: The block is blue and the block is a triangle.

     blue, a triangle

     1. T T T   2. T F F   3. F T F   4. F F F

     p ∧ q is false when either p or q is false.

C. The block is blue or a triangle.

     blue or a triangle or both blue and a triangle

1. T T T   2. T F T   3. F T T   4. F F F

p ∨ q is false when both p and q are false.

1.3B   Pages 21 and 22

I. A. If the block is blue, then it is a triangle.

B. blue, triangle

C. 1. T T T   2. T F F   3. F T T   4. F F T

D. p is true and q is false

II. A. The block is blue if and only if the block is a triangle.

If the block is blue, then it is a triangle and if the block is a triangle, then it is blue.

B. blue, triangle, triangle, blue

C. 1. T T T T T   2. T F F T F   3. F T T F F   4. F F T T T

D. Either p or q, but not both, are false.

1.3C   Page 23

I. a. If the number is greater than seven, then the number is greater than four.

b. If the number is greater than four, then the number is greater then seven.

c. If the number is not greater than seven, then the number is not greater than four.

d. If the number is not greater then four, then the number is not greater than seven.

II. a. True   b. False(Six is greater than four, but it is not greater than seven.)

c. False(Six is not greater than seven, but it is greater than four.   d. True

III. statement and contrapositive; converse and inverse;

1.3D   Pages 24 - 26

I. divisible by twelve, divisible by four; divisible by twelve; divisible by four

T T T T T; T F F F T; F T T F T; F F T F T

Never; Yes; By definition a tautology is always true.(The test of divisibility by twelve requires that a number which is divisible by twelve must be divisible by four and three.

II. divisible by twelve, divisible by four; not divisible by four; not divisible by twelve;

T T F F T F T;   T F F T F F T;   F T T F T F T;   F F T T T T T

Never, yes. By definition a tautology is always true. (The test of divisibility by twelve requires that a number which is not divisible by twelve cannot be divisible by four.)

III. divisible by twelve; divisible by four

divisible by four; divisible by two

divisible by twelve; divisible by two

T T T T T T T T;   T T F T F F F T;   T F T F T T F T;   T F F F T F F T;

F T T T T T T T;   F T F T F T F T;   F F T T T T T T;   F F F T T T T T;

Never; yes; By definition a tautology is a statement which is always true.

In order to be divisible by twelve, a number must be divisible by four.

In order to be divisible by four, a number must be divisible by two.

Therefore, if a number is divisible by twelve it must be divisible by two.

2.1A   Pages 31 and 32

Answers will vary.

2.1B   Pages 33 and 34

I. A - F Answers will vary.

   G. two

   H. units, 3^0 ;   groups of three, 3^1 ;   groups of nine, 3^2

   I. 27 or 3^3

   J. 1. 211_{three} ; $1 + 3 + 18 = 22$;   2. 15

II. A - F Answers will vary.

III. A. $1 + 28 + 98 = 127$

   B. 21

   C. 250

2.1C   Page 35

I. B. 1. 3   2. 3   3. 9   4. 3

   C. 1 flat 2 units

II. A - G Answers will vary.

2.1D   Page 36

I. B. 1. 10   2. 10   3. 100   4. 1000   5. 100   6. 10

II. A. Two flats three longs five units

   B. Four flats two units

   C. One block six units

III. Represent 530 as five flats and three longs.

   Represent 503 as five flats and three units.

   Child will see that 530 is larger. (You may need to trade the three longs for thirty units.)

2.2A   Pages 39 and 40

I. A. 20; 4(5)   B. 24; 3 + 3(7)   C. 15; 4 + 3 + 4(2)   D. 20; 45 − 5(5)

E. 49; $60 - 2 - 3(3)$ F. 7776; 6^5 G. 567; 7×3^4 H. 16; $2000 \div 5^3$

I. .028; $35000 \div 10 \div 50^3$

II. A. 1   B. 2   C. 8   D. 9   E. 81   F. 64   G. 1024   H. 625   I. 15,625

   J. 279,936 m^n is larger if $m < n$.

2.2B   Page 41

I. A. fives   B. five   C. trade for next largest piece   D. 0, 1, 2, 3, 4

   E. 5^4 5^3 5^2 5^1 5^0

      625   125   25   5   1

   1. 214_{five} 2. 1442_{five} 3. 31133_{five}

II. 1. 47 $\boxed{-}$ 27 $\boxed{=}$ 20 $\boxed{-}$ 9 $\boxed{=}$ $\boxed{=}$ 2 ; 122_{three}

   2. 509 $\boxed{-}$ 256 $\boxed{=}$ 253 $\boxed{-}$ 64 $\boxed{=}$ $\boxed{=}$ $\boxed{=}$ 61 $\boxed{-}$ 16 $\boxed{=}$ $\boxed{=}$ $\boxed{=}$ 13 $\boxed{-}$ 4

     $\boxed{=}$ $\boxed{=}$ $\boxed{=}$ 1 ; 13331_{four}

   3. 1263 $\boxed{-}$ 216 $\boxed{=}$ $\boxed{=}$ $\boxed{=}$ $\boxed{=}$ $\boxed{=}$ 183 $\boxed{-}$ 36 $\boxed{=}$ $\boxed{=}$ $\boxed{=}$ $\boxed{=}$ $\boxed{=}$ 3

     5503_{six}

   4. 2008 $\boxed{-}$ 343 $\boxed{=}$ $\boxed{=}$ $\boxed{=}$ $\boxed{=}$ $\boxed{=}$ 293 $\boxed{-}$ 49 $\boxed{=}$ $\boxed{=}$ $\boxed{=}$ $\boxed{=}$ 48

     $\boxed{-}$ 7 $\boxed{=}$ $\boxed{=}$ $\boxed{=}$ $\boxed{=}$ $\boxed{=}$ $\boxed{=}$ 6 ; 5566_{seven}

   5. 1328 $\boxed{-}$ 729 $\boxed{=}$ 599 $\boxed{-}$ 81 $\boxed{=}$ $\boxed{=}$ $\boxed{=}$ $\boxed{=}$ $\boxed{=}$ $\boxed{=}$ 32 $\boxed{-}$ 9

     $\boxed{=}$ $\boxed{=}$ $\boxed{=}$ 5 ; 1735_{nine}

2.2C   Pages 42 and 43

   1. 5 $\boxed{\times}$ 7 $\boxed{=}$ $\boxed{M+}$; 1 $\boxed{\times}$ 1 $\boxed{=}$ $\boxed{M+}$ $\boxed{MR}$; 36

   2. 3 $\boxed{\times}$ 4 $\boxed{=}$ $\boxed{M+}$; 2 $\boxed{\times}$ 1 $\boxed{=}$ $\boxed{M+}$ $\boxed{MR}$; 14

   3. 5 $\boxed{\times}$ 36 $\boxed{=}$ $\boxed{M+}$; 4 $\boxed{\times}$ 6 $\boxed{=}$ $\boxed{M+}$; 3 $\boxed{\times}$ 1 $\boxed{=}$ $\boxed{M+}$ $\boxed{MR}$; 207

   4. 1 $\boxed{\times}$ 125 $\boxed{=}$ $\boxed{M+}$; 1 $\boxed{\times}$ 5 $\boxed{=}$ $\boxed{M+}$; 3 $\boxed{\times}$ 1 $\boxed{=}$ $\boxed{M+}$ $\boxed{MR}$; 133

   5. 8 $\boxed{\times}$ 9 $\boxed{=}$ $\boxed{M+}$; 7 $\boxed{\times}$ 1 $\boxed{=}$ $\boxed{M+}$ $\boxed{MR}$; 79

   6. 4 $\boxed{\times}$ 25 $\boxed{=}$ $\boxed{M+}$; 2 $\boxed{\times}$ 1 $\boxed{=}$ $\boxed{M+}$ $\boxed{MR}$; 102

   7. 2 $\boxed{\times}$ 64 $\boxed{=}$ $\boxed{M+}$; 1 $\boxed{\times}$ 4 $\boxed{=}$ $\boxed{M+}$; 3 $\boxed{\times}$ 1 $\boxed{=}$ $\boxed{M+}$ $\boxed{MR}$; 135

   8. 5 $\boxed{\times}$ 36 $\boxed{=}$ $\boxed{M+}$; 4 $\boxed{\times}$ 6 $\boxed{=}$ $\boxed{M+}$; 3 $\boxed{\times}$ 1 $\boxed{=}$ $\boxed{M+}$ $\boxed{MR}$; 207

   9. 2 $\boxed{\times}$ 512 $\boxed{=}$ $\boxed{M+}$; 1 $\boxed{\times}$ 64 $\boxed{=}$ $\boxed{M+}$; 3 $\boxed{\times}$ 8 $\boxed{=}$ $\boxed{M+}$;

     7 $\boxed{\times}$ 1 $\boxed{=}$ $\boxed{M+}$ $\boxed{MR}$; 1119

10. 2 $\boxed{\times}$ 27 $\boxed{=}$ $\boxed{M+}$; 2 $\boxed{\times}$ 9 $\boxed{=}$ $\boxed{M+}$; 1 $\boxed{\times}$ 3 $\boxed{=}$ $\boxed{M+}$;

     1 $\boxed{\times}$ 1 $\boxed{=}$ $\boxed{M+}$ $\boxed{MR}$; 76

3.1A   Pages 49 and 50

I. 1. H   2. E   3. I   4. I   5. J

II. D - 4 ; E - 5 ; F - 6 ; G - 7 ; H - 8 ; I - 9 ; J -10 ;

III.A. D + C = G or 7

B. H + B = J or 10

IV.

	B	C	D	E	F	G	H	I	J	J+A
	C	D	E	F	G	H	I	J	J+A	J+B
	D	E	F	G	H	I	J	J+A	J+B	J+C
	E	F	G	H	I	J	J+A	J+B	J+C	J+D
	F	G	H	I	J	J+A	J+B	J+C	J+D	J+E
	G	H	I	J	J+A	J+B	J+C	J+D	J+E	J+F
	H	I	J	J+A	J+B	J+C	J+D	J+E	J+F	J+G
	I	J	J+A	J+B	J+C	J+D	J+E	J+F	J+G	J+H
	J	J+A	J+B	J+C	J+D	J+E	J+F	J+G	J+H	J+I
	J+A	J+B	J+C	J+D	J+E	J+F	J+G	J+H	J+I	2J

V. Yes. C + E = H and H = E + C. Therefore, C+E=E+C.

VI. Yes. B + (A + D)=B + E=G and (B + A) + D=C + D=G.

VII. No. The set has 10 rods, or 10 elements. There are no single rods for numbers > 10, so the set is not closed.

3.1B   Pages 51 − 54

I. A. 3. D

B. 1. J − H=10 − 8=2=B   2. F − D=6 − 4=2=B   3. G − B=7 − 2=5=E

4. I − C= 9 − 3=6=F   3. D − A=4 − 1=3=C

C.
```
      0  -  -  -  -  -  -  -  -  -
   A  O  -  -  -  -  -  -  -  -
   B  A  0  -  -  -  -  -  -  -
   C  B  A  0  -  -  -  -  -  -
   D  C  B  A  0  -  -  -  -  -
   E  D  C  B  A  0  -  -  -  -
   F  E  D  C  B  A  0  -  -  -
   G  F  E  D  C  B  A  0  -  -
   H  G  F  E  D  C  B  A  0  -
   I  H  G  F  E  D  C  B  A  0
```

D. No. H − E = C and E − H is undefined.

E. No. G − (E − B) = G − C=D.   (G − E) − B = B − B=0.

F. No. For example, D − G is undefined since the difference cannot be represented using centimeter rods.

II. A. The D rod.

B. 1. $H + \underline{B} = J$; $J - H = B$ 2. $D + \underline{B} = F$; $F - D = B$ 3. $B + \underline{E} = G$; $G - B = E$.

    4. $C + \underline{F} = I$; $I - C = F$.   5. $A + \underline{B} = D$; $D - A = B$

C. 1. H   2. C   3. E   4. 5

D. 1. B, G, E, 5, 5   2. D, I, E, 5, 5   3. F, J, D, 4, 4

E. 1. $7 - 5 = G - E$; $E + \underline{B} = G$. So $G - E = B$ or 2.

    2. $10 - 4 = J - D$; $D + \underline{F} = J$. So $J - D = F$ or 6.

3.2A   Page 57

1. BT: F-1   L-10   U-12 ;   AT: F-2   L-1   U-2

2. BT: F-7   L-15   U-10 ;   AT: F-8   L-6   U-0

3. BT: L-15   U-10 ;   AT: F-1   L-6   U-0

4. BT: F-15   L-12   U-11 ;   AT: B-1   F-6   L-3   U-1

5. BT: F-11   L-15   U-11 ;   AT: B-1   F-2   L-6   U-1

3.2B   Page 58

1. BW: L-4   U-8 ;   Difference: L-1   U-9

2. BW: F-1   L-0   U-3 ;   Difference: L-4   U-7

3. BW: F-2   L-2   U-4 ;   Difference: L-3   U-7

4. BW: B-2   F-0   L-0   U-1 ;   Difference: F-1   L-0   U-4

3.2C   Page 59

1. BT: L-3 U-2; AT: F-1 U-2;   2. BT: F-3 L-2 U-3;   AT: B-1 F-1;

3. BT: F-3 L-2 U-4; AT: B-1 F-1 U-1   4. BT: B-1 F-1 L-2 U-3; AT: B-1 F-2;

5. BT: B-1 F-1 L-3 U-3; AT: B-1 F-2 L-1   6. BT: B-1 F-1 U-3; AT; B-1 F-1 L-1

3.2D   Page 60

1. BW: L-2 U-1; DIF: U-2   2. BW: F-1 U-1; DIF: U-2

3. BW: F-2; DIF: F-1 L-1 U-2   4. BW: F-2 L-1 U-1; DIF: F-2 U-2

5. BW: B-1; DIF: F-1 U-2   6. BW: B-1 F-1 U-2; DIF: F-2 U-1

4.1A   Pages 65 and 66

Example: 1. D   2. H   3. 8 A rods   4. 8

I. A. 3 B rods are equal in length to one F rod. The F rod is 6 A rods long. $3 \times 2 = 6$.

B. 2 E rods are equal in length to one J rod. The J rod is 10 A rods long. $2 \times 5 = 10$.

C. 3 C rods are equal in length to one I rod. The I rod is 9 A rods long. $3 \times 3 = 9$.

II. A. 3 F rods are equal in length to one J rod and one H rod. The J and the H rods together are 18 A rods long. $3 \times 6 = 18$.

   B. 2 G rods are equal in length to one J rod and one D rod. The J and the D rods together are 14 A rods long. $2 \times 7 = 14$.

   C. 6 B rods are equal in length to one J rod and one B rod. The J and the B rods together are 12 A rods long. $6 \times 2 = 12$.

   D. 5 D rods are equal in length to two J rods. The J rods together are 20 A rods long. $5 \times 4 = 20$.

III. A. Yes. 2 C rods are equal in length to 3 B rods.

   B. Yes. Two trains of 6 C rods each are equal in length to three trains of 2 F rods each. (Note: $(2 \times 6) \times 3 = 3 \times (2 \times 6)$ by the commutative property.)

   C. Yes. $3(B + C) = 3 E = J + E$. $3 B + 3 C = F + I = J + E$.

4.1B   Pages 67 and 68

I. A. 3 rows of 2 tiles each or 6 tiles.

   B. 2 rows of 5 tiles each or 10 tiles.

   C. 3 rows of 3 tiles each or 9 tiles.

   D. 3 rows of 6 tiles each or 18 tiles.

   E. 2 rows of 7 tiles each or 14 tiles.

   F. 6 rows of 2 tiles each or 12 tiles.

   G. 5 rows of 4 tiles each or 20 tiles.

II. A. Yes. 2 rows of 3 tiles contain 6 tiles. 3 rows of 2 tiles contain 6 tiles. $2 \times 3 = 3 \times 2$.

   B. Yes. 2 arrays of 3 rows of 4 tiles together contain 24 tiles. 4 arrays of 2 rows of 3 tiles together contain 24 tiles.

   C. Yes. 3 arrays of 9 tiles together contain 27 tiles. 3 rows of 4 tiles added to 3 rows of 5 tiles together contain 27 tiles.

4.1C   Pages 69 and 70

I. A. Twelve square tiles may be sorted into 2 equal piles of 6 tiles each. $12 \div 2 = 6$.

   B. Twenty square tiles may be sorted into 4 equal piles of 5 tiles each. $20 \div 4 = 5$.

   C. Twenty-one tiles may be sorted into 3 equal piles of 7 tiles each. $21 \div 3 = 7$.

   D. Fifteen tiles may be sorted into 5 equal piles of 3 tiles each. $15 \div 5 = 3$.

II. A. Twenty tiles may be sorted into 6 equal piles of 3 tiles each with 2 tiles left over. $20 \div 6 = 3 \text{ R}2$.

   B. Twenty-six tiles may be sorted into 4 equal piles of 6 each with 2 left over. $26 \div 4 = 6 \text{ R}2$.

   C. Ten tiles may be sorted into 3 equal piles of 3 each with 1 left over. $10 \div 3 = 3 \text{ R}1$.

D. Twenty-seven tiles may be sorted into 5 equal piles of 5 each with 2 left over. $27 \div 5 = 5$ R2.

E. Twenty-three tiles may be sorted into 7 equal piles of 3 each with 2 left over. $23 \div 7 = 3$ R2.

F. Thirty tiles may be sorted into 4 equal piles of 7 each with 2 left over. $30 \div 4 = 7$ R2.

4.1D Pages 71 and 72

I. A. $4 \times n = 12$. We must build an array of 4 rows using 12 tiles. Each row has 3 tiles. $n = 3$.

B. $5 \times n = 15$. We must build an array of 5 rows using 15 tiles. Each row has 3 tiles. $n = 3$.

C. $3 \times n = 21$. We must build an array of 3 rows using 21 tiles. Each row has 7 tiles. $n = 7$.

D. $2 \times n = 18$. We must build an array of 2 rows using 18 tiles. Each row has 9 tiles. $n = 9$.

II. A. $n \times 3 = 6$. How many C rods are equal in length to the F rod? $n = 2$.

B. $n \times 2 = 8$. How many B rods are equal in length to the H rod? $n = 4$.

C. $n \times 4 = 12$. How many D rods are equal in length to the J rod plus the B rod? $n = 3$.

D. $n \times 9 = 18$. How many I rods are equal in length to the J rod plus the H rod? $n = 2$.

4.2 Pages 75 and 76

I. A. Answers may vary. Ex. $200 \times 380 = 76000$

B. 1.

2. $192 \times (300 + 70 + 6) = (192 \times 300) + (192 \times 70) +$ $(192 \times 6) = 57600 + 13440 + 1152 = 72192$

C. Yes. Both take into consideration the place value of each digit of the factors. Both take partial products and add them.

D. The standard algorithm also takes partial product and gives regard to the place value of each digit of the factors.

II. A. Answers may vary. Ex. $500 \div 20 = 25$.

B. 1.

√ 18	1
36	2
72	4
√ 144	8
√ 288	16
450	25

2.

18	450	10
	− 180	
	270	10
	− 180	
	90	5
	− 90	

C. Yes. Both algorithms concern multiples of the divisor. The Egyptian Method adds the multiples to arrive at the dividend. The Subtractive Method subtracts the multiples from the dividend to arrive at zero.

D. The standard algorithm is also concerned with subtracting multiples of the divisor from the dividend.

4.3A   Pages 79 and 80

1. a. 1×12, 2×6, 3×4;   b. 1, 2, 3, 4, 6, 12   c. $1 \mid 12$; $2 \mid 12$; $3 \mid 12$; $4 \mid 12$; $6 \mid 12$; $12 \mid 12$;

2. a. 1×8, 2×4; $1 \mid 8$; $2 \mid 8$; $4 \mid 8$; $8 \mid 8$;   b. 1×7; $1 \mid 7$; $7 \mid 7$;

   c. 1×20, 2×10, 4×5; $1 \mid 20$; $2 \mid 20$; $4 \mid 20$; $5 \mid 20$; $10 \mid 20$; $20 \mid 20$;

   d. 1×16, 2×8, 4×4; $1 \mid 16$; $2 \mid 16$; $4 \mid 16$; $8 \mid 16$; $16 \mid 16$;

3. a. 8, 20, 16   b. Answers will vary. Ex. 28 is composite. Rectangular arrays of the dimensions 1×28, 2×14, are possible 4×7.

4. a. 7   b. Answers may vary. Three is prime because the only possible rectangle has dimensions 1×3.

5. a. Fifteen tiles can be arranged in rectangles with dimensions of 1×15 and 3×5.
   Fourteen tiles can be arranged in rectangles with dimensions of 1×14 and 2×7.
   Since their only common dimension is 1 they are relatively prime.

   b. Eight tiles can be arranged in rectangles with dimensions of 1×8 and 2×4.
   Twelve tiles can arranged in rectangles with dimensions of 1×12, 2×6, and 3×4.
   Since they have a common dimension other than 1, they are not relatively prime.

   c. Answers will vary. Ex. Sixteen tiles may be arranged in rectangles with dimensions of 1×16, 2×8, and 4×4. Twenty-five tiles may be arranged in rectangles with dimensions of 1×25 and 5×5. Since they have no common dimension they are relatively prime.

4.3B   Page 81

The prime numbers between one and one hundred are 2, 3, 5, 7, 11, 13, 17, 19, 23, 29, 31, 37, 41, 43, 47, 53, 59, 61, 67, 71, 73, 79, 83, 89, and 97.

4.4A   Page 85

1. No. Reflexive property. Ex. The B rod is not two A rods shorter than itself.
   Symmetric property. Ex. The B rod is two A rods shorter than the D rod, but the D rod is not two A rods shorter than the B rod.
   Transitive property. Ex. The B rod is two A rods shorter than the D rod and the D rod is two A rods shorter than the F rod, but the B rod is not two A rods shorter than the F rod.

2. No. Reflexive property. Ex. The B rod is not longer than itself.
   Symmetric property. Ex. The B rod is longer than the A rod, but the A rod is not longer than the B rod.

3. No. Reflexive property. Ex. The B rod is not half as long as itself.

Symmetric property. Ex. The B rod is half as long as the D rod, but the D rod is not half as long as the B rod.

Transitive property. Ex. The B rod is half as long as the D rod and the D rod is half as long as the H rod, but the B rod is not half as long as the H rod.

4. Yes. {red blocks, yellow blocks, blue blocks}

5. No. Reflexive property. Ex. The small blue triangle is not a different color than itself.

Transitive property. Ex. The small blue triangle is a different color than the small red square and the small red square is a different color than the large blue circle, but the small blue triangle is not a different color than the large blue circle.

6. Yes. {triangles, squares, circles}

4.4B   Page 86 and 87

I. A − C. Answers will vary.

D. Yes. Each person is assigned exactly one day on which they were born. The domain is the set of classmates in C. The range is the set of days on which these people were born(not necessarily Sunday through Saturday, inclusive).

E. Not necessarily; only if each person was born on a different day of the week.

II. A. 1. The range of f is {1, 2, 3, 4, 5, ...}.

2. It means perform function f twice in succession using the first output as the second input. Not defined; Range is not a subset of the domain. Ex. Two will be the output when 4 is entered, but 2 cannot be used as an input because it is not in the domain.

B. 1. Range of g: {4, 16, 36, 64, 100, . . .}

2. It means perform function g twice in succession using the first output as the second input. It is defined because the range of g is a subset of the domain of g.

C. 1. Perform g followed by f in succession. It is defined because the range of g is a subset of the domain of f.

2. Perform f followed by g in succession. It is not defined because the range of f is not a subset of the domain of g.

4.4C   Pages 88 − 90

I. A.

2	3	4	5	6	7	8	9	10	11	12	1
3	4	5	6	7	8	9	10	11	12	1	2
4	5	6	7	8	9	10	11	12	1	2	3
5	6	7	8	9	10	11	12	1	2	3	4
6	7	8	9	10	11	12	1	2	3	4	5
7	8	9	10	11	12	1	2	3	4	5	6

B. 1. Yes. $3 \oplus 4 = 4 \oplus 3$

(Pattern in table.)

2. Yes. $(3 \oplus 6) \oplus 9 = 3 \oplus (6 \oplus 9)$

3. Yes. You always get a result between 1 and 12 , inclusive.

8	9	10	11	12	1	2	3	4	5	6	7
9	10	11	12	1	2	3	4	5	6	7	8
10	11	12	1	2	3	4	5	6	7	8	9
11	12	1	2	3	4	5	6	7	8	9	10
12	1	2	3	4	5	6	7	8	9	10	11
1	2	3	4	5	6	7	8	9	10	11	12

C. Yes. Twelve is identity element. $12 \oplus n = n \oplus 12 = n$. This will take you completely around the clock.

D. Yes. $1 \oplus 11=12$; $2 \oplus 10=12$; $3 \oplus 9=12$; $4 \oplus 8=12$; $5 \oplus 7=12$; $6 \oplus 6=12$; $7 \oplus 5=12$; $8 \oplus 4=12$; $9 \oplus 3=12$; $10 \oplus 2=12$; $11 \oplus 1=12$;

II. A. What do you add to 8 to get 4? The answer is 8.

B. Yes. The difference of each pair of numbers is defined. Every column and row contains all 12 digits.

III. A.

1	2	3	4	5	6	7	8	9	10	11	12
2	4	6	8	10	12	2	4	6	8	10	12
3	6	9	12	3	6	9	12	3	6	9	12
4	8	12	4	8	12	4	8	12	4	8	12
5	10	3	8	1	6	11	4	9	2	7	12
6	12	6	12	6	12	6	12	6	12	6	12
7	2	9	4	11	6	1	8	3	10	5	12
8	4	12	8	4	12	8	4	12	8	4	12
9	6	3	12	9	6	3	12	9	6	3	12
10	8	6	4	2	12	10	8	6	4	2	12
11	10	9	8	7	6	5	4	3	2	1	12
12	12	12	12	12	12	12	12	12	12	12	12

B. 1. Yes. Because of patterns in the table, $A \otimes B$ always equals $B \otimes A$.

2. Yes, Because of patterns in the table $A \otimes (B \otimes C)$ always equals $(A \otimes B) \otimes C$. Ex. $6 \otimes (4 \otimes 8)=(6 \otimes 4) \otimes 8$; $12=12$.

3. Yes. Every pair of numbers has one and one result under $\otimes$.

C. Yes. One because $1 \otimes n=n \otimes 1=n$.

D. No. Ex. $2 \otimes n = 1$. Taking trips of 2 will never get you to 1.

IV. A. $n=3$ because $5 \otimes 3 = s$.   B. $1 \oslash 2=n$; $2 \otimes n=1$. There is no value for N that makes this statement true. One does not appear in the column and row assigned to 2.

4.4D   **Pages 91 and 92**

I. A.

0	1	2
1	2	0
2	0	1

B. Tracy   C. Tracy   D. Tracy. Move the brad ten units.

E. Divide 11 by 3. The remainder is 2. Two represents Tracy.

II. A.

0	2	1
1	0	2
2	1	0

B. Tracy   C. If we know that Robert babysat today, then who babysat three days ago?

III. A.

0	0	0
0	1	2

B. 1. Yes. $A \times B = B \times A$

2. Yes. $A \times (B \times C)=(A \times B) \times C$.

0   2   1   3. Yes. Each pair of factors results in a product of 0, 1, or 2.

IV. $1 \div 2 = n$; $n \times 2 = 1$; How many "trips" of 2 would be taken to arrive at 1? $n = 2$.

5.1A   Pages 97 and 98

1. a. ⊖ ⊖ ⊖ ⊖ ⊕ ⊕ ⊕ ⊕   Result: 0   b. ⊕ ⊕ ⊕ ⊕ ⊖ ⊖ ⊖ ⊖   Result: 0

2. a. Yes   b. Yes. Equal numbers of positive and negative chips will cancel one another. In either case the result is zero.   c. Commutative property of addition.

3. a. ⊕ ⊕ ⊕ ⊖ ⊖ ⊖ ⊖ ⊖   Result: −2   b. ⊖ ⊖ ⊖ ⊖ ⊖ ⊖ ⊖ ⊖ ⊖   Result: −9

   c. ⊖ ⊖ ⊕ ⊕ ⊕ ⊕   Result: 2   d. ⊕ ⊕ ⊕ ⊕ ⊕ ⊕ ⊕ ⊕ ⊖   Result: 7

5.1B   Pages 99 − 102

I. A. 1. Remove 2 positive chips.   2. Add 2 positive and 2 negative chips.   3. 8 negative chips.

   4. ⊖ ⊖ ⊖ ⊖ ⊖ ⊖ + ⊕ ⊕ ⊖ ⊖ − ⊕ ⊕ = 8 negative chips or − 8.

B. 1. ⊖ ⊖ ⊖ ⊖ ⊖ ⊖ ⊖ + ⊕ ⊕ ⊕ ⊕ ⊖ ⊖ ⊖ ⊖ − ⊕ ⊕ ⊕ ⊕ = 11 negative chips or −11.

   2. ⊕ ⊕ ⊕ ⊕ ⊕ ⊕ ⊕ ⊕ − ⊕ ⊕ ⊕ = 5 positive chips or 5.

   3. ⊕ ⊕ + ⊕ ⊕ ⊖ ⊖ − ⊕ ⊕ ⊕ ⊕ = 2 negative chips or −2.

   4. ⊕ ⊕ ⊕ ⊕ ⊕ ⊕ ⊕ ⊕ ⊕ + ⊕ ⊖ − ⊖ = 10 positive chips or 10.

   5. ⊖ ⊖ ⊖ + ⊖ ⊖ ⊖ ⊕ ⊕ − ⊖ ⊖ ⊖ ⊖ ⊖ = 3 positive chips or 3.

   6. ⊖ ⊖ ⊖ ⊖ + ⊕ ⊕ ⊕ ⊕ ⊕ ⊖ ⊖ ⊖ ⊖ − ⊕ ⊕ ⊕ ⊕ ⊕ = 9 negative chips or −9.

II. 1. ⊖ ⊖ ⊖ ⊖ ⊖ ⊖ + ⊖ ⊖ = 8 negative chips or −8.

   2. ⊖ ⊖ ⊖ ⊖ ⊖ ⊖ ⊖ + ⊖ ⊖ ⊖ ⊖ = 11 negative chips or −11.

   3. ⊕ ⊕ ⊕ ⊕ ⊕ ⊕ ⊕ ⊕ + ⊕ ⊕ ⊕ = 11 positive chips or 11.

   4. ⊕ ⊕ + ⊖ ⊖ ⊖ ⊖ = 2 negative chips or −2.

   5. ⊕ ⊕ ⊕ ⊕ ⊕ ⊕ ⊕ ⊕ ⊕ + ⊕ = 10 positive chips or 10.

   6. ⊖ ⊖ ⊖ + ⊕ ⊕ ⊕ ⊕ ⊕ ⊕ = 3 positive chips or 3.

III. 1. $+2 + n = -6$. What must be added to 2 positive chips to have 6 negative chips?   −8.

   2. $+4 + n = -7$. What must be added to 4 positive chips to have 7 negative chips?   −11.

   3. $+3 + n = +8$. What must be added to 3 positive chips to have 8 positive chips?   5.

   4. $+4 + n = +2$. What must be added to 4 positive chips to have 2 positive chips?   −2.

   5. $-1 + n = +9$. What must be added to 1 negative chip to have 9 positive chips?   10.

   6. $-6 + n = -3$. What must be added to 6 negative chips to have 3 negative chips?   3.

5.2A   Pages 105 and 106

1. a. $(-3) + (-3) + (-3) + (-3)$ b. Add 4 groups of 3 negative chips.

   c. ⊖ ⊖ ⊖ + ⊖ ⊖ ⊖ + ⊖ ⊖ ⊖ + ⊖ ⊖ ⊖ ;   −12

2. a. No   b. Subtraction   c. No   d. 12 positive and 12 negative chips   e. −12.

f. $\ominus \ominus \ominus \ominus \ominus \ominus \ominus \ominus \ominus \ominus \ominus \ominus \ - \ \oplus \oplus \oplus \ - \ \oplus \oplus \oplus \ - \ \oplus \oplus \oplus \ - \ \oplus \oplus \oplus$

3. Begin with 12 positive chips and 12 negative chips. Remove 4 groups of 3 negative chips. Twelve postive chips remain.

4. a. Add three groups of 5 negative chips. Result: -15.

   b. Begin with 6 positive and 6 negative chips. Remove 6 "groups" of 1 negative chip. Result: $+6$.

   c. Begin with 16 positive and 16 negative chips. Remove 8 groups of 2 negative chips. Result: $+16$.

   d. Begin with 8 positive and 8 negative chips. Remove 2 groups of 4 positive chips. Result: -16.

   e. Add 7 groups of 2 positive chips. Result: $+14$.

   f. Begin with 10 positive and 10 negative chips. Remove 2 groups of 5 positive chips. Result: -10.

5.2B   Pages 107 and 108

1. a. $-2 \times n = +8$ b. subtract   c. 2   d. $+8$

   e. Begin with a name for zero which includes 8 positive chips. (8 positive and 8 negative chips)

   f. $\oplus \oplus \oplus \oplus \oplus \oplus \oplus \oplus \ominus \ominus \ominus \ominus \ominus \ominus \ominus \ominus \ - \ \ominus \ominus \ominus \ominus \ - \ \ominus \ominus \ominus \ominus$

   g. 4   h. -4 i. -4.

2. a. $\oplus \oplus + \oplus \oplus + \oplus \oplus$; $+2$ b. $\oplus \oplus \oplus \oplus \ominus \ominus \ominus \ominus \ - \ \ominus \ominus \ - \ \ominus \ominus$; -2

   c. $\oplus \oplus \oplus \oplus \oplus \oplus \oplus \oplus \oplus \oplus \oplus \oplus \ominus \ominus \ominus \ominus \ominus \ominus \ominus \ominus \ominus \ominus \ominus \ominus \ - \ \oplus \oplus \oplus \oplus \ -$

   $\oplus \oplus \oplus \oplus \ - \ \oplus \oplus \oplus \oplus$; $+4$ d. $\ominus \ominus + \ominus \ominus + \ominus \ominus + \ominus \ominus + \ominus \ominus$; -2

   e. $\oplus \oplus \oplus \oplus \oplus \oplus \oplus \oplus \ominus \ominus \ominus \ominus \ominus \ominus \ominus \ominus \ - \oplus \oplus \ - \ \oplus \oplus \ - \ \oplus \oplus \ - \ \oplus \oplus$; $+2$

   f. $\oplus \oplus \oplus \oplus \oplus \oplus \ominus \ominus \ominus \ominus \ominus \ominus \ - \ \ominus \ominus \ominus \ - \ \ominus \ominus \ominus$; -3

3. No. $[(-16) \div (+4)] \div -2 \ \neq \ (-16) \div [(+4) \div (-2)]$.

4. No. $(+8) \div (-4) \neq (-4) \div (+8)$.

5.3   Pages 111 and 112

A. 1. x represents the number and kind of chips in set X which must be added to 3 positive chips in order to have a set of 4 positive chips. We remove 3 positive chips from x + 3 and from 4. x must be 1 positive chip, or $+1$.

   2. x represents the number and kind of chips in set X which must be added to 2 negative chips to get 4 negative chips. We remove 2 negative chips from x + (−2) and from −4. x must be 2 negative chips, or -2.

   3. x represents the number and kind of chips which when duplicated will give us 6 negative chips. Divide 2x and −6 by 2. x must represent 3 negative chips.

   4. x represents the number and kind of chips which can be divided into 4 equal groups of 2 negative chips each. Multiply x ÷ 4 by 4 and −2 by 4. x must represent 4 groups of −2, or -8.

   5. x represents the number and kind of chips which we can duplicate and then add 4 positive chips

to for a result of 12 positive chips. First subtract 4 positive chips from $2x + 4$ and from 12. $2x$ must equal 8. Two groups of x chips represent 8 so one group must represent 4 positive chips, or $+4$.

B. 1. One more than 3 times a set of x chips is more than 7 positive chips. Subtract 1 positive chip from $3x + 1$ and from 7. Three groups of x chips must be greater than 6. Since 3 groups of 2 positive chips is equal to 6, we know that x must be greater than 2.

2. Four more than a group of x chips is greater than 9 positive chips. Remove 4 chips from $x + 4$ and from 9. x must be greater than 5.

3. x represents a set we can double and then remove 3 positive chips from to get less than 5 positive chips. First add 3 positive chips to $2x - 3$ and to 5. Then $2x$ must be less than 8. Therefore, x must be less than 4.

6.1A   Pages 117 and 118

II. $\frac{1}{8}$, $\frac{1}{6}$, $\frac{1}{4}$, $\frac{1}{3}$, $\frac{1}{2}$; less

III.A. 6   B. 9 ; Show 9 twelfths exactly cover 3 fourths.

IV. A. 4   B. 8; Show 8 twelfths exactly cover 2 thirds.

V. A. $\frac{1}{4}$ B. $\frac{1}{2}$ C. $\frac{1}{2}$ D. $\frac{1}{2}$

VI. A. $\frac{1}{2}$ B. 1   C. 2   D. 1

VII.A. $\frac{1}{2}$ B. 2   C. 1   D. 1

6.1B   Pages 119 and 120

I. A. 1. a. 12   b. $\frac{1}{12}$ 2. a. 6   b. $\frac{1}{6}$ 3. a. 4   b. $\frac{1}{4}$

4. a. 3   b. $\frac{1}{3}$ 5. a. 2   b. $\frac{1}{2}$

B. $\frac{1}{12}$, $\frac{1}{6}$, $\frac{1}{4}$, $\frac{1}{3}$, $\frac{1}{2}$; As the denominator increases, the fraction decreases, if the numerator is held constant.

C. Five B rods are longer than 3 C rods.

D. Two D rods are longer than 5 A rods.

II. A. 1. $\frac{1}{18}$ 2. $\frac{1}{9}$ 3. $\frac{1}{6}$ 4. $\frac{1}{3}$ 5. $\frac{1}{2}$

B. $\frac{1}{18}$, $\frac{1}{9}$, $\frac{1}{6}$, $\frac{1}{3}$, $\frac{1}{2}$

C. 1. Two B rods are shorter than 1 F rod.   2. Three A rods are equal in length to 1 C rod.

3. One I rod is longer than 3 B rods.

III.A. 1. $\frac{1}{24}$　　　2. $\frac{1}{12}$　　　3. $\frac{1}{8}$　　　4. $\frac{1}{6}$　　　5. $\frac{1}{4}$　　　6. $\frac{1}{3}$

B. 1. Two H rods are longer than 5 C rods.　　2. Three C rods are shorter than 5 B rods.

3. Five D rods are shorter than 7 C rods.

6.1C　　Pages 121 and 122

I. A. An infinite number.

B. Answers may vary. For example, $\frac{5}{20}$ and $\frac{4}{20}$.

C. Answers may vary.

D. $\frac{250}{1000}$, $\frac{200}{1000}$

E. Answers may vary. For example, $\frac{203}{1000}$.

F. Yes.

G. Answers may vary.

II. Answers may vary. Examples are given below.

1. $\frac{21}{140}$　　　2. $\frac{11}{42}$　　　3. $\frac{17}{144}$　　　4. $\frac{115}{1320}$　　　5. $\frac{302}{9300}$

III.A. $\frac{2}{9}$　　　B. $\frac{2}{9}$　　　C. $\frac{3}{13}$　　　D. $\frac{3}{13}$

IV. 1. $\frac{4}{24}$, or $\frac{1}{8}$　　　2. $\frac{3}{13}$　　　3. $\frac{2}{17}$　　　4. $\frac{2}{23}$　　　5. $\frac{2}{61}$

V. No. There are an infinite number of fractions between any pair of fractions.

6.1D　　Page 123

1. Four F rods are equal in length to 3 H rods. LCM is 24.

2. Five D rods are equal in length to 2 J rods. LCM is 20.

3. Five C rods are equal in length to 3 E rods. LCM is 15.

4. Three H rods are equal in length to 8 C rods. LCM is 24.

5. Seven C rods are equal in length to 3 G rods. LCM is 21.

6. Six D rods, 3 H rods, and 8 C rods are equal in length. LCM is 24.

6.2A　　Page 127

I. A. 12 A rods, 6 B rods, 4 C rods, 3 D rods, 2 F rods,

B. A$\frac{1}{12}$; B$\frac{1}{6}$; C$\frac{1}{4}$; D$\frac{1}{3}$; F$\frac{1}{2}$;

C. We can replace the D rod with two B rods. Then we add 5 B rods to the 2 B rods. These are equal in length to 1 J rod and 2 B rods, or $1\frac{1}{6}$.

D. We replace three C rods with nine A rods. We then add the 9 A rods to the 1 A rod for a total of 10 A rods. These are equal in length to 5 B rods, or $\frac{5}{6}$.

II. We could let the J rod represent one whole. The A rod would then represent $\frac{1}{10}$ and the B rod would represent $\frac{1}{5}$. We then replace three B rods with six A rods and add them to the given A rod for a total of seven tenths, or $\frac{7}{10}$.

III. We could use two J rods as one whole. The A rod would then represent $\frac{1}{20}$ and the D rod would represent $\frac{1}{5}$. We add one A rod to four D rods. This train is equal to 17 A rods, or $\frac{17}{20}$.

6.2B   Page 128

I. 1. Let the J rod and the B rod represent one whole. Five B rods will represent five sixths. Seven A rods will represent seven twelfths. The C rod can be added to the seven A rods to form a train equal in length to the five B rods. Since four C rods are the same length as the J and B rod, the C rod represents $\frac{1}{4}$ and is the difference.

   2. Let the J rod and the B rod represent one whole. Three C rods will represent three fourths. The D rod will represent one third. Five A rods may be added to 1 D rod to get 3 C rods. The difference is $\frac{5}{12}$.

   3. Let the J rod and the B rod represent one whole. One A rod may be added to one F rod to get 7 A rods. The difference is $\frac{1}{12}$.

II. Let the J and the D rods represent one whole. Compare the length of 2 J and 2 D rods together with the length of one B rod. One J rod, one D rod, and 6 B rods may be added to 1 B rod to get a length equal to 2 J and 2 D rods. The difference is $1\frac{6}{7}$.

6.3A   Page 131

I. 1. Three F rods represent $1\frac{1}{2}$.   2. Six C rods represent $1\frac{1}{2}$.   3. Sixteen A rods represent $1\frac{1}{3}$.

II. $2\frac{1}{3}$ may be represented as 2 J rods, 2 B rods, and 1 D rod. $4 \times 2\frac{1}{3}$ means 8 J rods, 8 B rods, and 4 D rods. 8 J and 8 D rods represent 8. 4 D rods represent $1\frac{1}{3}$. Therefore, $9\frac{1}{3}$ is the result.

6.3B   Page 132

1. Fold in half horizontally and shade 1 part. Fold into fourths vertically and shade 3 parts. Three of eight parts have been shaded twice. Therefore, the answer is $\frac{3}{8}$.

2. Fold into four parts horizontally and shade 1 part. Fold into thirds vertically and shade 2 parts. Two of 12 parts have been shaded twice. Therefore, the answer is $\frac{2}{12}$.

3. Fold into eight parts horizontally and shade 3 parts. Fold in half vertically and shade 1 part. Three of 16 parts have been shaded twice. Therefore, the answer is $\frac{3}{16}$.

4. Fold 4 rectangles into thirds and shade 2 parts of each. $\frac{8}{3}$ have been shaded.

5. Begin with 3 rectangles folded in half horizontally. Shade both parts of 2 rectangles and 1 part of the other. Fold each one into thirds vertically and shade 2 parts. Each rectangle has been

divided into 6 parts. Ten of these sixths have been shaded twice. Therefore, the answer is $\frac{10}{6}$.

6.3C   Pages 133 and 134

1. What fraction of two D rods is the F rod? Using A rods we see that the F rod is $\frac{6}{8}$, or $\frac{3}{4}$, of the 2 D rods. The result, therefore, is $\frac{3}{4}$.

2. How many F rods are equal in length to 3 C rods? The answer is $1\frac{1}{2}$.

3. What fraction of 5 B rods is 7 A rods? Using additional A rods, we see that the 5 B rods are equal in length to 10 A rods. The 7 A rods are, therefore, $\frac{7}{10}$ of the 5 B rods.

4. What fraction of 3 C rods is one B rod? The 3 C rods are 9 A rods in length and the B rod is 2 A rods in length. Therefore, the B rod is $\frac{2}{9}$ of the 3 C rods.

5. How many C rods are equal in length to 5 A rods? The result is $1\frac{2}{3}$.

6. How many groups of 3 C rods are equal in length to the J and B rods? The result is $1\frac{1}{3}$.

7. How many groups of 2 D rods are equal in length to 2 J and 2 B rods? The result is 3.

8. What fraction of 2 J and 2 B rods is 2 D rods? Six D rods are equal in length to 2 J rods and 2 B rods. Therefore, 2 D rods are $\frac{1}{3}$ of 2.

9. How many A rods are equal in length to the J rod, the B rod, and the D rod? (The D rod is $\frac{1}{3}$.) The result is 16.

6.3D   Page 135

1. a. First fold a rectangle in half horizontally and shade one section. Then fold the rectangle in thirds vertically and shade 2 sections. Two of 16 sections have been shaded twice. Therefore, the result is $\frac{2}{6}$, or $\frac{1}{3}$.

   Next fold another rectangle in thirds horizontally and shade two sections. Then fold it in half vertically and shade one section. Two of 16 sections have been shaded twice. Therefore, the result is also $\frac{2}{6}$, or $\frac{1}{3}$.

   b. Begin with 2 D rods. One D rod is half the length. The result is $\frac{1}{2}$.

   The result is the same if you begin with an F rod. Two B rods represent two thirds. The 2 B rods are $\frac{1}{3}$ of the J and the B rods.

2. How many C rods are equal in length to 2 D rods? The result is $2\frac{2}{3}$.

   The result is NOT the same if we ask, "How much of 2 D rods is the C rod?" Using A rods, we see the result is $\frac{3}{8}$.

3. a. Fold a rectangle in thirds horizontally and shade 1 section of one rectangle. This is one third. Multiplying by two means that we want two such sections. Shade the second one. Then fold the rectangle into fourths vertically and shade 1 section. The result is two sections out of 12 shaded twice. This is the same as $\frac{1}{6}$.

   The result is the same is we fold a rectangle into fourths horizontally and shade one section and fold it into thirds vertically and shade one section. This will result in one shaded section out of twelve which is $\frac{1}{12}$. Two twelfths is the same as $\frac{1}{6}$.

b. The I rod would represent one third of 2 J and 2 B rods. One fourth of this rod would be the B rod. The B rod represents $\frac{1}{6}$ of the J and the B rods.

The result is the same is one fourth of the D rod, which represents one third. One fourth is an A rod. Two A rods are equal in length to a B rod. The B rod represents $\frac{1}{6}$ of the J and B rods.

6.4A   Page 139

Answers will vary.

6.4B   Pages 140 − 142

I. A. $\frac{5}{2}$ B. 1. $\frac{2}{7}$ 2. $\frac{2}{5}$ 3. $\frac{7}{2}$ 4. $\frac{5}{7}$ 5. $\frac{7}{5}$ C. $\frac{1}{6}$, $\frac{6}{1}$, $\frac{1}{7}$, $\frac{7}{1}$, $\frac{6}{7}$, $\frac{7}{6}$

II. A. 1. a. 2   b. 6   c. $\frac{15}{6} = \frac{5}{2}$; 30 = 30 ; Yes.   e. 21   f. Answers vary.(ex. $\frac{15}{21}$)

2. a. 3   b. 20   c. $\frac{8}{20}$ d. $\frac{8}{20} = \frac{2}{5}$; 40 = 40. Yes.   e. 28   f. Answers vary.(ex.$\frac{20}{28}$)

B. 1. $\dfrac{\text{number of pieces with no right angles}}{\text{total number of tangram pieces}} = \frac{1}{7} = \frac{4}{n}$; n = 28.

2. $\dfrac{\text{number of pieces with at least one right angle}}{\text{total number of tangram pieces}} = \frac{6}{7} = \frac{n}{21}$; 7n = 126 ; n = 18.

6.4C   Page 143

Answers will vary.

7.1   Page 149

I. 10, 100, 1000; $\frac{1}{10}$, $\frac{1}{100}$, $\frac{1}{1000}$; .1, .01, .001

II. A. 1 block, 3 flats, 4 longs   B. 6 flats, 5 longs, 4 units   C. 3 units   D. 2 flats, 3 units

III. .305 may be represented as 3 flats and 5 units. .350 may be represented as 3 flats and 5 longs.

.350 is greater because 5 longs represent 50 units.

IV. A. 10, .1   B. 100, .01   C. 10, 10

7.2A   Page 150

A. Six longs and 8 units are added to 1 flat, 3 longs, and 5 units. The result is 1 flat, 9 longs, and 13 units. Ten units are traded for 1 long resulting in 1 flat, 10 longs, and 3 units. The ten longs are traded for 1 flat which results in 2 flats and 3 units, or .203.

B. Seven units are added to 9 flats, 9 longs, and 4 units. The result is 9 flats, 9 longs, and 11 units. We trade 10 units for 1 long giving us 9 flats, 10 longs, and 1 unit. We trade 10 longs for 1 flat giving us 10 flats and 1 unit. We trade 10 flats for 1 block giving us 1 block and 1 unit, or 1.001.

C. Nine flats and seven units are added to 3 flats and 6 longs. The result is 12 flats, 6 longs, and 7 units. We trade 10 flats for a block. The result is 1 block, 2 flats, 6 longs, and 7 units, or 1.267.

D. Three flats and 7 longs are added to seven flats, five longs, and two units. The result is 10 flats, 12 longs, and 2 units. We trade 10 longs for a flat giving us 11 flats, 2 longs, and 2 units. Then we trade 10 flats for a block giving us 1 block, 1 flat, 2 longs, and 2 units, or 1.122.

7.2B   Page 151

A. Begin with 6 flats, 2 longs, and 8 units. We must remove 5 units. Three units remain. We cannot remove 3 longs from 2 longs. We trade one flat for ten longs and remove 3 longs. Nine longs remain. We remove 4 flats from the 5 flats which remain which leaves us 1 flat. The result is 1 flat, 9 longs, and 3 units, or .193.

B. Begin with 8 flats, 5 longs, and 3 units. Because we cannot remove 8 units from 3 units, we trade 1 long for 10 units and remove the 8 units. Five units remain. We cannot remove 6 longs from 4 longs. We trade 1 flat for 10 longs and then remove 6 longs. Eight longs remain. Next we remove 5 flats from 7 flats. We now have 2 flats. The result is 2 flats, 8 longs, and 5 units, or .285.

C. Begin with 8 flats and 2 longs. We must remove 4 units. We trade 1 long for 10 units. We remove the 4 units. Six units remain. The result is 8 flats, 1 long, and 6 units, or .816.

D. Begin with 1 block and 3 units. We cannot remove 7 units from 3 units. We have no longs and no flats to trade. We must trade the block for 10 flats and one of those flats for 10 longs. Now we can trade 1 long for 10 units. Removing 7 units leaves us with 6 units. The result is 9 flats, 9 longs, and 6 units, or .996.

7.3   Page 152

1. 100%   2. a. 4   b. $\frac{1}{4}$ c. 25%   d. 25%   3. a. $\frac{8}{16} = \frac{1}{2}$; 50%   b. $\frac{4}{16} = \frac{1}{4}$; 25%
c. $\frac{12}{16} = \frac{3}{4}$; 75%   d. $\frac{4}{16} = \frac{1}{4}$; 25%   4. $37\frac{1}{2}\%$ is half of 75%. 75% of 16 squares is 12 squares. $37\frac{1}{2}\%$ of 16 squares must be 6 squares. Shade any 6 squares on the geoboard.

7.4   Pages 155 and 156

I. centimeters; decimeters   centimeters; decimeter

II. millimeters; millimeter, centimeter, decimeter

III.C. millimeters   decimeter

   centimeters   centimeter

   decimeters   millimeter

IV. Answers will vary.

V. Answers will vary.

7.5 A   Pages 157 and 158

1. Form a right triangle with legs of lengths 2 units and 1 unit. The hypotenuse has length $\sqrt{5}$. irrational.

2. Form a right triangle with legs of lengths 3 units and 2 units. The hypotenuse has length $\sqrt{13}$. irrational.

3. Form a right triangle with legs of lengths 4 units and 3 units. The hypotenuse has length $\sqrt{25}$, or 5. rational

4. We can construct a line segment with an irrational length, c, by drawing a right triangle with legs of lengths a and b, where $a^2 + b^2 = c^2$. The hypotenuse will have length c.

Answers to #5 − 8 will vary.

5. We draw a right triangle with legs of 3 and 2. The hypotenuse has length $\sqrt{13}$.

6. We draw a right triangle with legs of 5 and 7. The hypotenuse has length $\sqrt{74}$.

7. Use any length line segment as one unit. Construct a right triangle with legs both equal to 1. The hypotenuse has length $\sqrt{2}$. Construct another right triangle with legs 1 and $\sqrt{2}$ by copying the hypotnuse and the unit lengths. The hypotenuse has length $\sqrt{3}$. Construct third triangle with legs $\sqrt{3}$ and 2 by copying the hypotenuse of length $\sqrt{3}$ and 2 unit lengths. The hypotenuse has length $\sqrt{7}$.

8. Use any line segment as 1 unit. Construct a right triangle with legs of lengths 1 and 4. The hypotenuse has length $\sqrt{5}$. Construct a right triangle with legs 4 and $\sqrt{5}$. The hypotenuse has length $\sqrt{21}$.

7.5B   Page 159

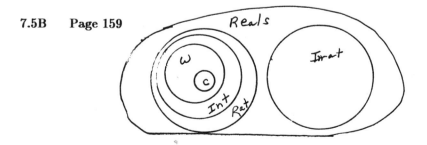

8.1A   Pages 165 and 166

Answers will vary.

8.1B   Pages 167 and 168

Answers will vary.

8.1C   Page 169

I − III Answers will vary.

IV. The graph in Part III gives the impression of less differences among the birthdays.

8.1D   Pages 170 − 172

I − VII Answers will vary.

VIII. Graph VII B − More dramatic difference.   Graph VII A − Less dramatic difference.

IX. To emphasize change in a line graph decrease the number of units on the vertical scale.

To de-emphasize change in a line graph increase the number of units on the vertical scale.

8.1E   Page 173

Answers will vary.

8.1F   Pages 174 and 175

I and II Answers will vary.

III. A. 5 and $\frac{1}{2}$ symbols   B. 6 and $\frac{23}{25}$ symbols   C. 6 and $\frac{4}{5}$ symbols

IV. A. $\frac{33}{50}$;   a little more than half a symbol   B. 5 and $\frac{22}{25}$; which is 5 and approximately $\frac{9}{10}$ symbols   C. 3 and $\frac{43}{75}$; which 3 symbols and a little more than half of another symbol.

8.2A   Page 179

Answers will vary.

8.2B   Pages 180 and 181

Answers will vary.

8.2C   Page 182

Answers will vary.

9.1A   Page 189 and 190

I. A. Point up, point down   B. No, not equally likely, varies with different tacks.

C. 5. Answers will vary.   6. Answers will vary.   7. Because we tossed 10 tacks ten times. This is the same as tossing 1 tack 100 times.   8. Answers will vary.

9. $\frac{100}{100}$ or 1; complementary

II. A. Top, bottom, side   B. No. Not equally likely; depends upon dimensions of can.

C. 3. Answers will vary.   4. Answers will vary.   5. $\frac{30}{30}$ or 1; Sum of probabilities should be one because one of them is certain to occur.

9.1B   Page 191

I. A. P(yellow block) or P(blue block) or P(block that is both yellow and blue)

B. P(red block) or P(triangle) or P(red triangle)

II. A. $\frac{6}{18}$ or $\frac{1}{3}$ B. $\frac{6}{18}$ or $\frac{1}{3}$ C. 0, No blocks are blue and yellow.

III. Yes because there are no blocks that are blue and yellow.

IV. A. $\frac{6}{18}$ or $\frac{1}{3}$ B. $\frac{6}{18}$ or $\frac{1}{3}$ C. $\frac{2}{18}$ or $\frac{1}{9}$

V. No because there are some red triangles. P(R or T) = P(R) + P(T) $-$ P(red traingle)

VI. P(A $\cup$ B) = P(A or B) = P(A) + P(B) $-$ P(A and B)

9.1C   Pages 192 and 193

I. A.
2	3	4	5	6	7
3	4	5	6	7	8
4	5	6	7	8	9
5	6	7	8	9	10
6	7	8	9	10	11
7	8	9	10	11	12

B. 11; 36   C. 1 2 3 4 5 6 5 4 3 2 1

D. 10

II. D. Answers will vary.   E. No, because of the limited sampling. As the number of trials gets very large, the relative frequencies will get close to the probabilities.

F. They should be closer. If we use results from entire class, they should be even closer.

G. Experimental, theoretical

9.2A   Pages 197 and 198

I. A.

Tree diagram

die	coin
	H
1	T
	H
2	T
	H
3	T
	H

Sample space

1H 2H 3H 4H 5H 6H 1T 2T 3T 4T 5T 6T

Note: The coin and die columns could be reversed.

B. $\frac{1}{12}$ C. 1, It is certain that one of the events will occur.

```
        4  <  T
              H
     5  <     T
              H
        6  <     T
```

II. A. 1. 1H 2H 3H 4H 5H 6H   2. 1   3. $\frac{1}{6}$ B. 1. 2H 2T 4H 4T 6H 6T   2. 3   3. $\frac{3}{6}$ or $\frac{1}{2}$

C. 1. $\frac{4}{6}$ or $\frac{2}{3}$ 2. $\frac{2}{6}$ or $\frac{1}{3}$ 3. $\frac{4}{8}$ or $\frac{1}{2}$ 4. $\frac{2}{4}$ or $\frac{1}{2}$ D. Not necessarily; C1 $\neq$ C3; C2 $\neq$ C4

E. Yes. $P(A \mid B) = P(A) = \frac{1}{2}$

9.2B   Pages 199 and 200

I. A.

1	2	3	4	5	6
2	4	6	8	10	12
3	6	9	12	15	18
4	8	12	16	20	24
5	10	15	20	25	30
6	12	18	24	30	36

B. 1. $\frac{9}{36}$ or $\frac{1}{4}$ 2. $\frac{27}{36}$ or $\frac{3}{4}$ C. −3.75

D. No. There is the danger of losing $3.75 each time.

E. −1.25

II. A.

2	3	4	5	6	7
3	4	5	6	7	8
4	5	6	7	8	9
5	6	7	8	9	10
6	7	8	9	10	11
7	8	9	10	11	12

B. 1. $\frac{18}{36}$ or $\frac{1}{2}$ 2. $\frac{18}{36}$ or $\frac{1}{2}$ C. −1.50

D. No. Lose average of $1.50 each time.

E. 0

9.3A   Pages 203 − 205

I. A. H T   1 1   B. HH HT TH TT   1 2 1

C. HHH HHT HTH THH TTH THT HTT TTT   1 3 3 1

D.

```
                        1
                     1     1
                  1     2     1
               1     3     3     1
            1     4     6     4     1
         1     5    10    10     5     1
      1     6    15    20    15     6     1
   1     7    21    35    35    21     7     1
1     8    28    56    70    56    28     8     1
1  9   36   84   126   126   84   36   9   1
```

E. The numerators of the probability of the possible outcomes when 3 coins are tossed; indicates the number of coins; F. 8; The number of outcomes in your sample space; G. 2^7

H. 1 7 21 35 35 21 7   Note: In Part H, each place indicates the number of ways in which to get 0, 1, 2, 3, 4, 5, 6, and 7 H, respectively, on 7 coins.   I. 1. $\frac{15}{64}$ 2. $\frac{70}{256}$ 3. $\frac{10}{32}$

J. Use the row in which the second entry is 7. One represents the number of ways to get one girl, 21 represents the number of ways to get 2 girls, 35 represents the number of ways to get 3 girls, 35 represents the number of ways to get 4 girls P(4 girls in family of 7 children) = $\frac{35}{128}$.

A-F. Answers may vary

II. G. Experimental, theoretical

9.3B   Pages 206 − 208

I.   **first   second   third**

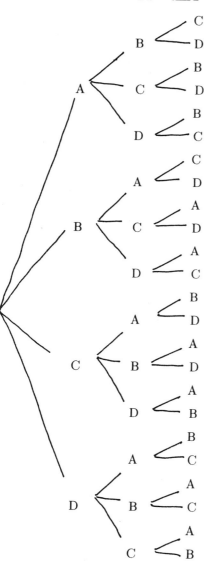

ABC ABD ACB ACD ADB ADC

BAC BAD BCA BCD BDA BDC

CAB CAD CBA CBD CDA CDB

DAB DAC DBA DBC DCA DCB

A. 1. 24   2. $4 \cdot 3 \cdot 2 = 24$ 3. $6 \cdot 5 \cdot 4 \cdot 3 = 360$

B. 1. 6   3. $4 \cdot 3 \cdot 2 \cdot 1 = 24$

C. 1. $\frac{24}{6} = 4$ Equal

D. 1. AB BA CA DA EA   2. 20; Equal
   AC BC CB DB EB   3. 2; Equal
   AD BD CD DC EC   4. $\frac{20}{2} = 10$
   AE BE CE DE ED   5. Equal

E. 1. Make a tree diagram with 4 columns.

   List the entire sample space.

   Group the arrangments which contain the same letters.

   Determine the number of such groups and divide by the number of arrangements in each group.

2. $\dfrac{\text{permutations of 10 things taken 4 at a time}}{\text{permutations of 4 things taken 4 at a time}} = \dfrac{10 \cdot 9 \cdot 8 \cdot 7}{4 \cdot 3 \cdot 2} = 210$

10.1A   Page 213

I. D.

   E. Answers may vary. Ex. A set of points equidistant from a set point.

II. A.

   B. 1. Line segment from center to point on the circle.

      2. Line segment connecting any two points on the circle.

      3. Chord which passes through the center of the circle.

      4. Distance around the circle, or its "perimeter".

10.1B   Pages 214 and 215

Answers will vary. Examples follow.

1.   Impossible.   2.

3.   4.

5.   6. Impossible.

10.1C   Pages 216 and 217

Outcomes vary.

10.1D   Pages 218 through 220

I. Square, parallelogram, five isosceles right triangles.   II – VII. Answers will vary.

10.2A   Pages 223 and 224

Construct models.

10.2B   Pages 225 and 226

I. Construct models.

II. A. Two parallel bases, sides are parallelograms.   B. One base, sides are triangles.

III. Yes. The tetrahedron is a triangular pyramid because it has one triangular base and the other sides are triangles.. The hexahedron is a square prism because it has two square bases which are parallel and the sides are parallelograms.

IV. A. 4, 4, 6   B. 8, 6, 12   C. 6, 8, 12   D. 15, 12, 25   E. 12, 20, 30
    F. 4, 4, 6   G. 5, 5, 8   H. 6, 5, 9   I. 8, 6, 12   J. 8, 6, 12

V. Yes. The sum of the number of vertices and the number of faces is equal to two more than the number of edges.

VI. V + F = E + 2

VII. A. 58   B. No, this is impossible because it would have only three faces and a polyhedron must have at least four faces.

VIII. No, the relationship does not hold for cones, cylinders, spheres, and hemispheres. Because of their curved edges the faces are not polygons. They are not polyhedra.

10.2C   Pages 227 and 228

I. B. Circle   D. Triangle   F. Ellipse   H. Parabola   J. Hyperbola

II. 1. Circle - every slice   2. Circle - slice parallel to bases; Rectangle - slice through bases; Ellipse — slice through side at angle without slicing bases;   3. Triangle - slice three faces; slice off a "corner"; Square - Slice parallel to a base; Rectangle - slice upper edge and opposite bottom edge; Trapezoid - slice four faces entering just in from a top vertex, at an angle, exiting through bottom; Pentagon - slice five faces; Hexagon - slice six faces; Quadrilateral - Slice any four faces;   4. Triangle - slice three sides; Quadrilateral - slice all four sides.

11.1A   Page 233

PRIMARY: obtuse, 120° ; INTERMEDIATE: acute, 30° ; JR. HIGH: obtuse, 110° ;
SR. HIGH: obtuse, 100° .

11.1B   Pages 234 and 235

I. C. Straight angle, 180°   D. 180°   II. C. 360°   D. 360°

III. Rhombus: 4, 2, 360°   Hexagon: 6, 4, 720°   Regular pentagon: 5, 3, 540°
    Pentagon: 5, 3, 540° The number of triangles formed is two less than the number of sides. The sum of the vertex angles could be represented as 180(n − 2).

11.1C   Page 236

I. A line segment which connects the midpoint of a side with the opposite vertex.

II. perpendicular bisector   III. Construct medians   IV. They are concurrent. (All three

medians intersect at the same point.)

11.1D   Pages 237 and 238

I. Construct rhombus.   II. 1. It is a parallelogram with four congruent sides.

2. Construct angle bisectors.   3. They are the diagonals.   4. Perpendicular

5. Each diagonal is cut into two congruent segments, but the segments from different

diagonals are not necessarily congruent.   III. Four congruent sides; Diagonals are perpendicular

to one another;   Diagonals bisect the angles ;   Diagonals bisect one another.

11.1E   Page 239

II. 1. $\dfrac{AF}{FG} = \dfrac{(AE + EF)}{FG} = \dfrac{10 + \sqrt{10^2 + 20^2}}{20} = \dfrac{10 + 10\sqrt{5}}{20} = \dfrac{(1 + \sqrt{5})}{2}$ 2. Approximately 1.618.

III. Some of the ratios of length to width probably approximate the golden ratio. Perhaps this is

because it is pleasing to the eye.

11.2A   Page 243

1. a.   b.   c.   d.   e.   f.   g.   h.   i.

II. No.   The distance between two nails diagonally is greater than 1 unit.   By the Pythagorean

Theorem

it is $\sqrt{2}$. Therefore the perimeter is $10 + 2\sqrt{2}$.

11.2B   Page 244

Your ratios should be approximately 3.14, but because of the difficulty in obtaining precise

measurements, particularly for small objects, your ratios probably differ from 3.14 by varying

amounts.

11.2C   Pages 245 and 246

I. 1. 7 square units   2. 5 square units   II. B. 29, 57, 43   C. 141, 199, 170

D. They are congruent.   E. 4   F. 35.25, 49.75, 42.5   G. They are very close, but those in

part F are more accurate because they unit used is smaller.   H. We could divide the squares into

smaller units.

11.2D   Pages 247 and 248

I. A. 2 3 6; 1 5 5; 3 4 12; 5 2 10; 4 2 8   B. A = length $\times$ width = lw

II. A. 1, 1; 2 4; 3 9; 4 16; 5 25   B. A = side × side or s^2

III. C. length (or width)   D. width (or length)   E. A = base × height or bh

IV. B. Yes   D. A = bh   E. one half   F. A = $\frac{1}{2}$bh

V. C. $A_1 = \frac{1}{2} b_1 h$ $A_2 = \frac{1}{2} b_2 h$ D. $\frac{1}{2} b_1 h + \frac{1}{2} b_2 h$ E. A = $\frac{1}{2} (b_1 + b_2) h$

VI. E. width   F. length   G. $\frac{1}{2} C × r$ H. A = πr^2

11.2E   Page 249

I. C. 1. 3   2. 3, 3, 9   3. 9   D. 1. 36   2. 36, 36, 1296   E. 9, 1296

II. C. 1. 10   2. 10, 10, 100   3. 100   D. 1. 100   2. 100, 100, 10000   3. 10000

E. 100, 10000

III. A. 144 ; Draw a square which is 1 ft on each side. Since there are 12 inches in a foot the
dimensions of the square could be written as 12 inches by 12 inches. The area of the
square is 12 × 12 or 144 square inches.

B. 100; Draw a square which is 1 dm on each side. Since there are 10 cm in 1 dm the
dimensions of the square could be written as 10 cm by 10 cm. THe area of the square
is 10 × 10 or 100 square centimeters.

11.2F   Pages 250 and 251

I. a. 6 square units   b. 5 square units

II. 3. 3 square units   a. 4 square units   b. 2 square units   c. 1 square unit   d. 8 square units

III. A. 4.5 square units   B. 4 square units   C. 6 square units   D. 6 square units   E. 6.5 square units
A = $\frac{BP}{2}$ + IP − 1

11.2G   Page 252

I. 1- Square prism (cube)   96 square units   96 square units

2- Square pyramid   108 square units   108 square units

3- Rectangular prism   80 square units   80 square units

II. Answers will vary.

11.3A   Page 255

I and II. Answers will vary.

III. Mass and weight are not synonymous. Mass is dependent upon the molecular structure of the
object and is constant, unless you are traveling at the speed of light. Weight is dependent

upon the pull of gravity on the object and , therefore, will vary at different locations on Earth and on other planets.

11.3B   Page 256

1-5. For water at room temperature: $1 \text{ dm}^3 = 1 \text{ l} = 1 \text{ kg}$.

6. Your number of cubic decimaters, liters, and kilograms will be equal to the number of people in your class.

11.3C   Page 257

I. and II. Answers will vary.   III. A. Navy beans.   B. The navy beans; because they are smaller they more closely follow the shape of the can and pack together more tightly.

IV. Answers will vary.   V. Volume is determined by the number of units it takes to fill the object. Cubic units facilitate the notion of filling because they are three-dimensional.

11.3D   Page 258

II. D. Three   E. $\frac{1}{3}$ F. $=; =$ G. $\frac{1}{3}$ H. Answers will vary.

III. D. Three   E. $\frac{1}{3}$ F. $=; =$ G. $\frac{1}{3}$ H. Answers will vary.

11.3E   Page 259

11.3F   Pages 260 and 261

5. One fourth of the peel will approximately cover the great circle.

11. You could use the string to measure the circumference of a great circle by marking off the distance and then comapring it to a ruler. You could then divide this number by and approximation for π and divide in half to get the radius. This is based upon the fact, $C = 2\pi r$.

11.3G   Pages 262 and 263

I. A. 32°, 0°;   98.6°, 37;   212° , 100°;   B. Comparing the two measures for normal body temperature, we know that 35° C would be rather warm for a sweater.

II. A. 1. 51.8°   2. 78.8°   3. 17.6°   4. 107.6°   5. 134.6°

B. 1. 52°; $+.2$ 2. 82°; $+3.2$ 3. 14°; -3.6 4. 114°; $+6.4$ 5. 144°; $+9.4$

C. Nine fifths is approximately equal to 2 and 32 is approximately equal to 30.

III. A. 1. $-1.1°$ 2. 16.7°   3. 48.9°   4. 35°   5. $-20°$

B. $C = \frac{1}{2}(F - 30)$; 1. 0°; $+1.1°$ 2. 16°; $-.7$ 3. 45°; -3.9 4. 32.5°; -2.5 5. $-17°$; $+3$

12.1A   Pages 269 and 270

I. C. 1. $10 - 3 = 7$ 2. $6 - 2 = 4$ D. $\sqrt{65}$

II. B. 2. (x_2, y_1) 3. $y_2 - y_1$ 4. $x_2 - x_1$ 5. $\sqrt{(x_2 - x_1)^2 + (y_2 - y_1)^2}$

   6. Same as number 5.

III. 3. Answers will vary.   4. 5.4; 7.8; 13.0   5. approximately 26.2 units

12.1B   Pages 271 and 272

I. A. 1. parallel   2. 3   3. Parallel lines have equal slopes.

   B. 1. perpendicular   2. $4; -\frac{1}{4}$ 3. Perpendicular lines have slopes which are negative reciprocals.

II. B. Square   C. $-\frac{11}{5}; \frac{5}{11}$ They are negative reciprocals. Perpendicular lines and right angles.

   D. Diagonals of a square intersect at right angles.

III. A. Rectangle   B. Two sides have slopes of $\frac{2}{1}$ and two sides have slopes of $-\frac{1}{2}$.

   C. Because both pair of opposite sides have the same slope they are parallel. A trapezoid

   has only one pair of parallel sides.

   D. Because the line segments intersecting at each vertex have slopes which are negative reciprocals

   the angle at each vertex is right. Therefore, it is a rectangle.

12.1C   Page 273

I. Each equation represents a parabola which has been shifted horizontally.

   The graph would be a parabola which has been shifted 7 units to the left.

II. Each equation represents a parabola which has been shifted vertically.

   The graph would be a parabola which has been shifted 6 units downward.

III. Each parabola has been shifted both horizontally and vertically, but all of them have the same

   shape and orientation.

   The graph would be a parabola which has been shifted 1 unit to the left and 8 units down.

IV. The graph is a parabola which has been shifted h units horizontally and k units vertically.

12.1D   Pages 274 and 275

I. B. (h, k) C. (x, y) D. $r = \sqrt{(x - h)^2 + (y - k)^2}$ E. $r^2 = (x - h)^2 + (y - k)^2$

II. A. 1. $(2, -4)$ 2. 5   3. Sketch   4. Answers will vary. Ex. $(7, -4), (2, -9), (-3, -4), (2, 1)$

   5. Each distance should be 5.

   B. 1. $\frac{4}{3}$ 2. $-\frac{3}{4}$ 3. Perpendicular lines because their slopes are negative reciprocals.

   4. $y = -\frac{3}{4}x + \frac{15}{4}$ or $3x + 4y = 15$.

12.2A   Page 279 – 281

I. 4. They are both 80°.   5. a. 4, 8, $\frac{1}{2}$ b. 3.5, 7, $\frac{1}{2}$ c. 2.5, 5, $\frac{1}{2}$ 6. Yes

7. Yes because the the corresponding angles are equal and the corresponding sides are proportional.   8. two angles, two angles, are similar

II. 1. a. 4   b. 2.4   c. 3.5   2. a. 12   b. 7.5   c. 10.5

5. a. 35° b. 65° c. 80° d. 35° e. 65° f. 80°

6. Corresponding angles are equal.   7. Yes. Corresponding angles are equal and corresponding sides are proportional.   8. three sides, proportional, three sides

III. 5. a. 80°   b. 50°   c. 80°   d. 50°   e. 10.4 cm   f. 5.2 cm

6. Yes. The cooresponding angles are equal and the corresponding sides are in proportion.

7. proportional; equal

12.2B   Pages 282 and 283

1. 2AG = PQ; 2GF = PR; etc.

2. Yes, ratio 1:2

3. Yes; corresponding angles equal corresponding sides in proportion.

12.2C   Pages 284 − 287

I. 3. a. 3 cm   b. 2.5 cm   c. 5 cm   d. 3 cm   e. 2.5 cm   f. 5 cm (order may vary.)

4. a. 120°   b. 32°   c. 28°   d. 120°   e. 32°   f. 28° (order may vary.)

5. The corresponding angles are equal.

6. Yes. They are exactly the same size and shape.

7. equal in length (congruent)

II. 4. a. 77°   b. 38°   c. 38°   d. 77°

5. The corresponding angles are equal.

6. a. 3.6 cm   b. 3.6 cm

7. Yes. Corresponding angles are equal and corresponding sides are equal.

8. two sides and the included angle of one are equal ; two sides and the included angle

III. 5. a. 77°   b. 77°

6. a. 2.4 cm   b. 3.5 cm   c. 2.4 cm   d. 3.5 cm

7. Yes. Corresponding sides are in proportion and corresponding angles are equal.

8. Two angles and the included side ; two angles and the included side

12.2D   Page 288

Answers will vary.

12.3A   Pages 291 — 293

A. 1.   FD 35   REPEAT 3 [FD 35 RT 120]   2. FD 30   REPEAT 5 [FD 30 RT 72]

        RT 120   RT 72

        FD 35   FD 30

        RT 120   RT 72

        FD 35   FD 30

        RT 120   RT 72

                                                                FD 30

                                                               RT 72

                                                                FD 30

                                                                RT 72

   3.   FD 25   REPEAT 6 [FD 25 RT 60]   4. FD 25   REPEAT 8 [FD 25 RT 45]

        RT 60   RT 45

        FD 25   FD 25

        RT 60   RT 45

        FD 25   FD 25

        RT 60   RT 45

        FD 25   FD 25

        RT 60   RT 45

        FD 25   FD 25

        RT 60   RT 45

        FD 25   FD 25

        RT 60   RT 45

                                                                FD 25

                                                               RT 45

                                                                FD 25

                                                                RT 45

B. The length of the side, the number of sides, the measure of an exterior angle.

C. REPEAT n [FD x RT $\frac{360}{n}$]   D. It looks more like a circle.

E. Answers will vary.   F. Answers will vary.   G. Answers will vary.

12.3B   Page 294 — 296

Example:

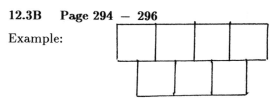

1. A. TO HEXAGON   B. TO EQTRIANGLE   C. TO OCTAGON

REPEAT 6 [FD 30 RT 60]   REPEAT 3 [FD 30 RT 120]   REPEAT 8 [FD 30 RT 45]

END   END   END

2. A. REPEAT 3 [SQUARE RT 90 FD 50 LT 90]

    PU

    LT 90

    FD 100

    RT 90

    BK 50

    REPEAT 3 [SQUARE RT 90 FD 50 LT 90]

  B. RT 30

    REPEAT 3 [EQTRIANGLE RT 60 FD 30 LT 90]

    PU

    LT 30

    FD 30

    LT 120

    FD 90

    RT 90

    REPEAT 3 [EQTRIANGLE RT 60 FD 30 LT 90]

  C. REPEAT 3 [HEXAGON1 RT 30 FD 30 RT 60 FD 30 LT 90]

    PU

    RT 150

    FD 30

    RT 60

    FD 30

    RT 60

    FD 30

    LT 60

    FD 30 RT 180 REPEAT 3 [HEXAGON1 RT 30 FD 30 RT 60 FD 30 LT 90]

12.3C   Page 297 and 298

I. B. A square of side 40.   C. No. The area will be 80^2 rather than 40^2 , or 6400 rather than 1600 square units. We demonstrate this by typing SQUARE 40 and then SQUARE 80.

  D. No. The area will be 20^2 rather than 40^2, or 400 rather than 1600 square units. We demonstrate this by typing SQUARE 40 and then SQUARE 20.

  E. The are a will be s^4 rather than s^2 . Type SQUARE 1600 and then SQUARE 40.

  F. Multiply by 3.

II. B. A rectangle of dimensions 50 and 100.   C. A = 2w × 2l = 4wl; The area is multiplied by 4.

Type RECTANGLE 100 200 and then RECTANGLE 50 100.

D. A = $\frac{1}{2}$ w × l = $\frac{1}{2}$ w l ; The area is divided by two. Type RECTANGLE 50 100 and then RECTANGLE 25 100.

E. A vertical line.

12.3D   Pages 299 and 300

I. It will draw square over itself.

II. A. A square   B. Squares rotated 90° each time.   C. Squares rotated 45° each time.

D.

III. Answers will vary.

A. TO TRIANGLE1   TO TRIANGLE2

RT 30   TRIANGLE1

REPEAT 3 [FD 30 RT 120]   RT 150

LT 30   TRIANGLE1

END   END

B. TO RECTANGLE1   TO RECTANGLE2

REPEAT 2 [FD 50 RT 90 FD 30 RT 90] RECTANGLE1

RT 90

RECTANGLE1

END   END

C. TO HEXAGON1   TO HEXAGON2

RT 30   HEXAGON1

REPEAT 6 [FD 30 RT 60]   RT 150

LT 30   HEXAGON1

END   END

13.1A   Page 305

5. Each distance is equal to the length of ray XY, which is approximately 4 cm.

6. They are congruent.   7. Move each vertex point three units to the right.

13.1B   Page 306

5. Line ℓ is the perpendicular bisector of the line segment connecting each point with its image.

6. They are congruent.

7. Measure the distance from each point to the line of reflection. Find the image of each point by extending the perpendicular from the point to the line and marking off an equal segment on the other side of the line of reflection.

13.1C   Pages 307 and 308

8. They are equal in measure..   9. They are equal in measure

11.a. Draw the line segments connecting each vertex with point P. Trace the figure and point P. Turn the traced copy clockwise using P as the point of rotation. Measure the angle of rotation at each vertex with your compass.

    b. Draw the line segments connecting each vertex with point P. Trace the figure and point P. Use your compass and straightedge to construct a sixty degree angle. It will be the measure of one of the angles of an equilateral triangle. Copy this angle of rotation at each vertex.

13.1D   Pages 309 and 310

3. The images do coincide. This will not always be true.

5. When the line of reflection and the ray of translation are parallel the image will be the same regardless of the order in which you perform them.

13.1E   Page 311

1 & 2:

Yes, 6;	No	Yes, 5
Yes, 1;	Yes, 3	No
No	Yes, 4	Yes, 8

3. The line of symmetry is a line of reflection.

13.1F   Page 312

1 & 2:

Yes, 2	No	Yes, 3
No	Yes, 5	No
Yes, 4	Yes, 2	No

3. We could consider the point of rotation as being located within the figure.

13.1G   Pages 313 and 314

I. The following figures tessellate: 1, 2, 3, 4, 5, 6, 7, 8, 10

II. A. 1. (3, 3, 3, 3, 3, 3)   2. (3, 3, 3, 3, 3, 3)   3. (3, 3, 3, 3, 3, 3)

4. (4, 4, 4, 4)   5. (4, 4 , 4, 4)   6. (4, 4, 4, 4)

7. (4, 4, 4, 4)   8. (4, 4, 4, 4)   10. (6, 6, 6)

13.1H   Pages 315 and 316

Answers will vary.

13.2A   Pages 319 and 320

I.

II.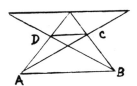

III. The greater the distance the greater the "height" of the picture.

13.2B   Page 321

I. 1,1;   2, 2;   3, 3;

II. Answers will vary. Examples follow.

   ring, compact disc; scissors, jacket, mask;   light switch plate, pretzel;

13.2C   Page 322

I. 4, 0, No; 8, 0, No; 0, 6, Yes; 15, 0, No; 12, 0, No.

II. No. Traversable only if only even vertices or exactly 2 odd vertices.

APPENDIX 2

MANIPULATIVES

Area Grid A

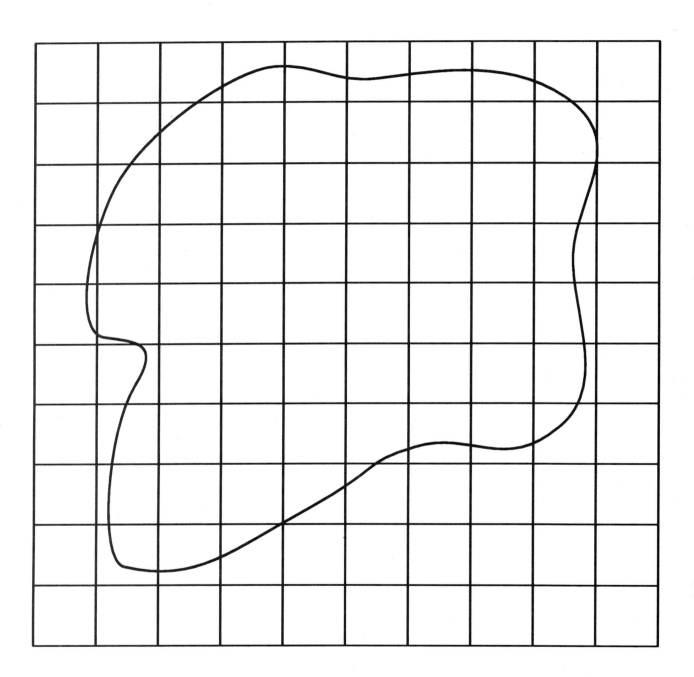

Area Grid B

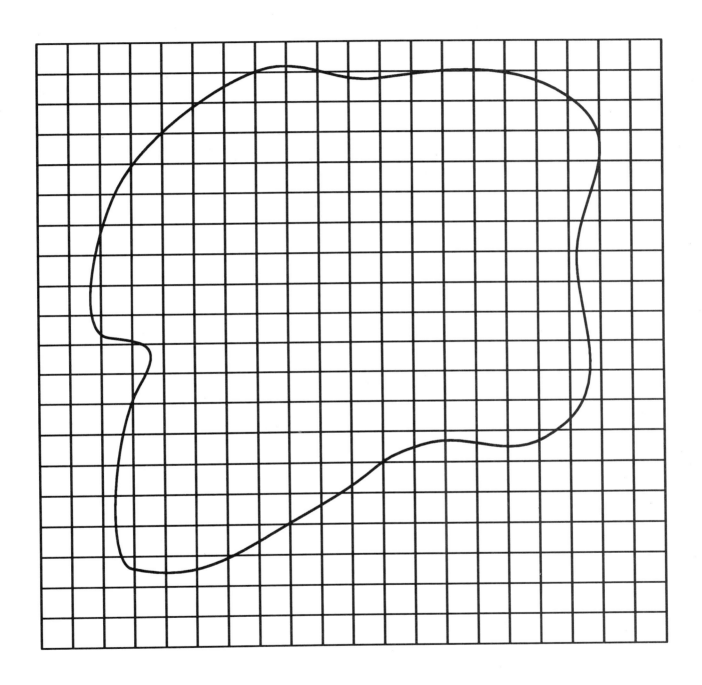

Area Grids (11.2D)

I. Rectangles

1. (rectangle grid 3×2)

4. (rectangle grid 5×2)

2. (rectangle grid 5×1)

5. (rectangle grid 4×2)

3. (rectangle grid 4×3)

II. Squares

1. (square 1×1)

2. (square 2×2)

3. (square 3×3)

4. (square 4×4)

5. (square 5×5)

III. Parallelogram

IV. Triangles

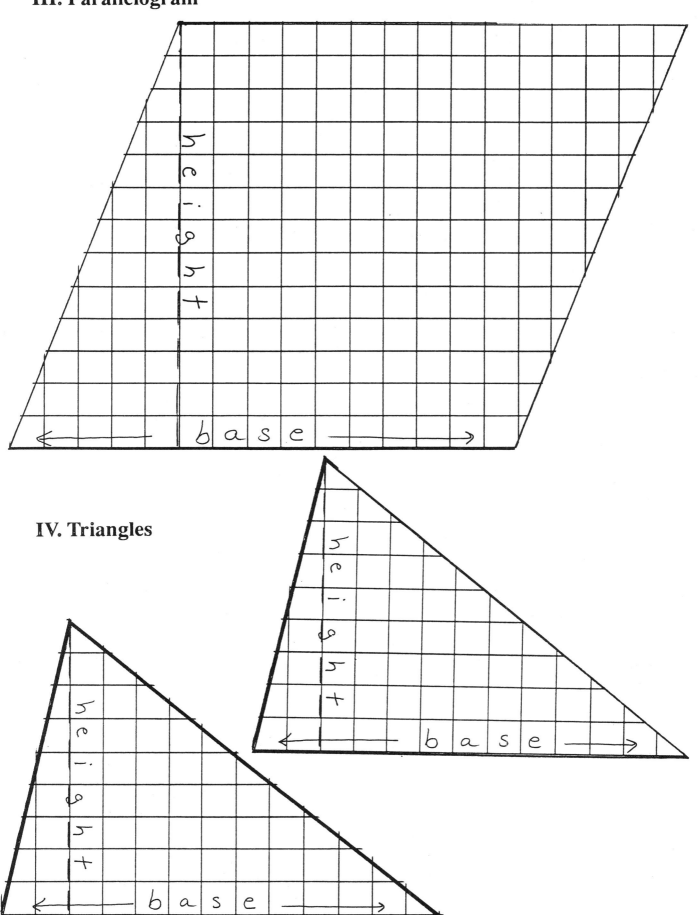

V. Trapezoid

base 1

base 2

VI. Circle

Circumference

RADIUS

APPENDIX 3

CUT-OUT MANIPULATIVES

Chapter 1 – Pattern Blocks

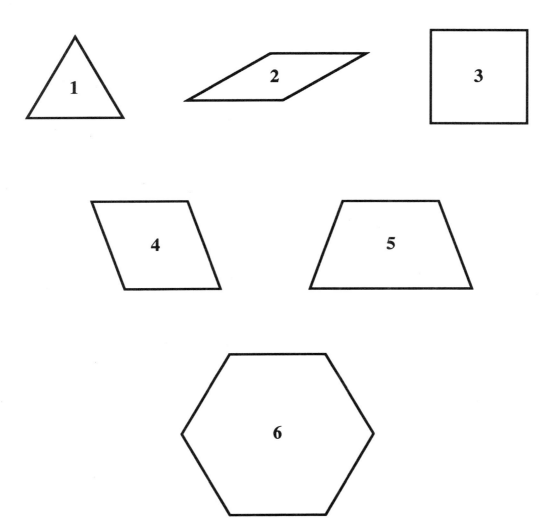

Chapter 1, 4, 9 – Attribute Blocks

R, r = red
B, b = blue
Y, y = yellow

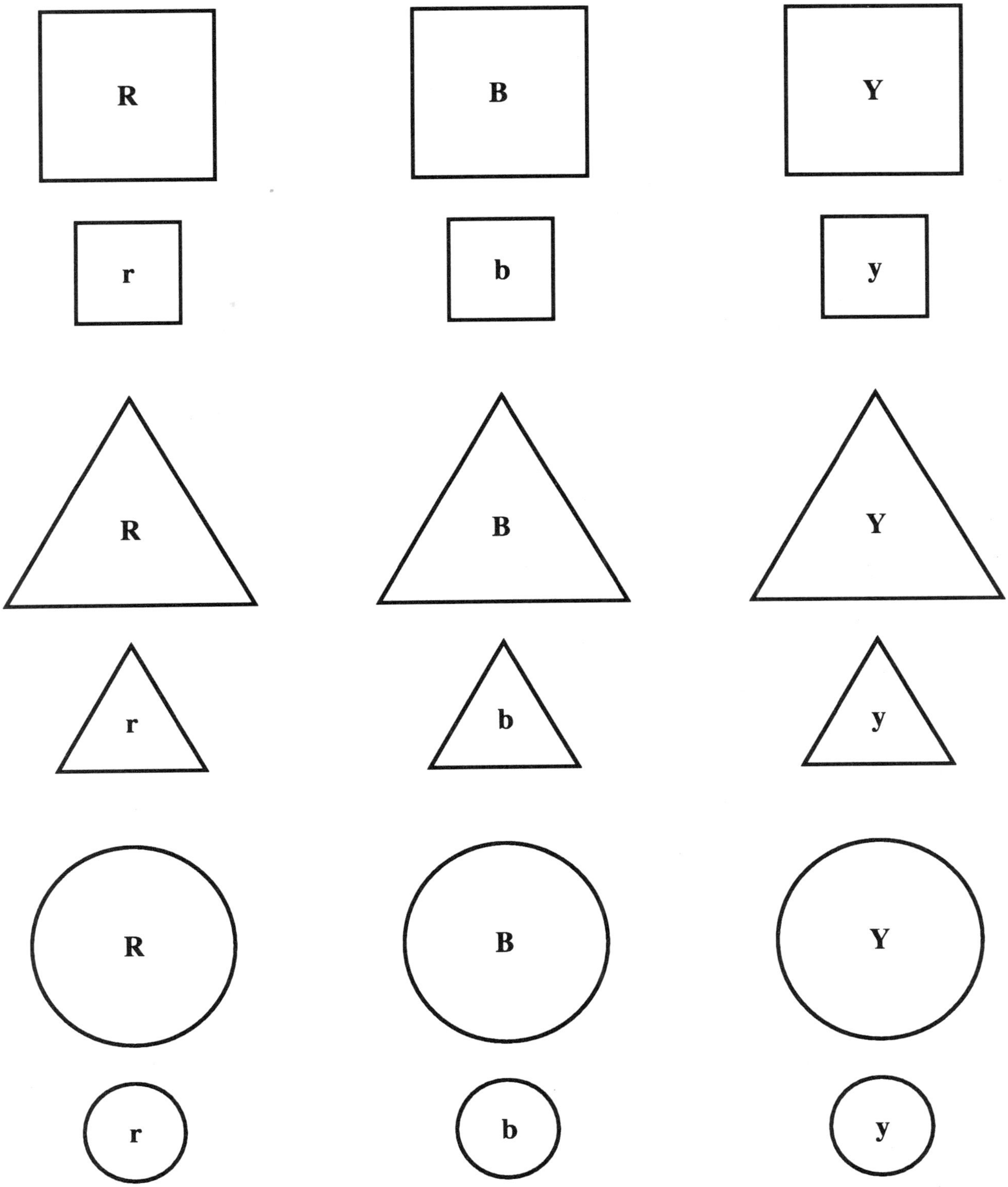

R (large square)	B (large square)	Y (large square)
r (small square)	b (small square)	y (small square)
R (large triangle)	B (large triangle)	Y (large triangle)
r (small triangle)	b (small triangle)	y (small triangle)
R (large circle)	B (large circle)	Y (large circle)
r (small circle)	b (small circle)	y (small circle)

Chapter 2 – Chip Trading Board

R	W	B

Chapters 2,3 – Base Three Pieces

Chapters 2,3,7 – Base Ten Pieces

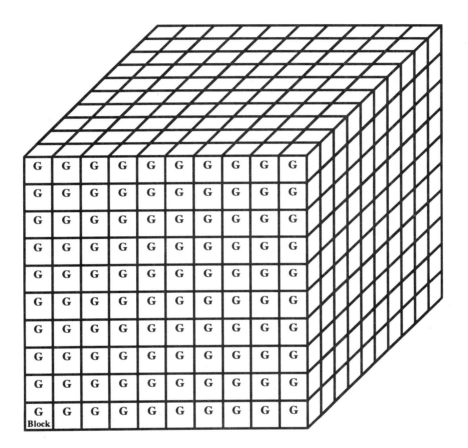

G Block

Chapters 2,3,7 – Base Ten Pieces

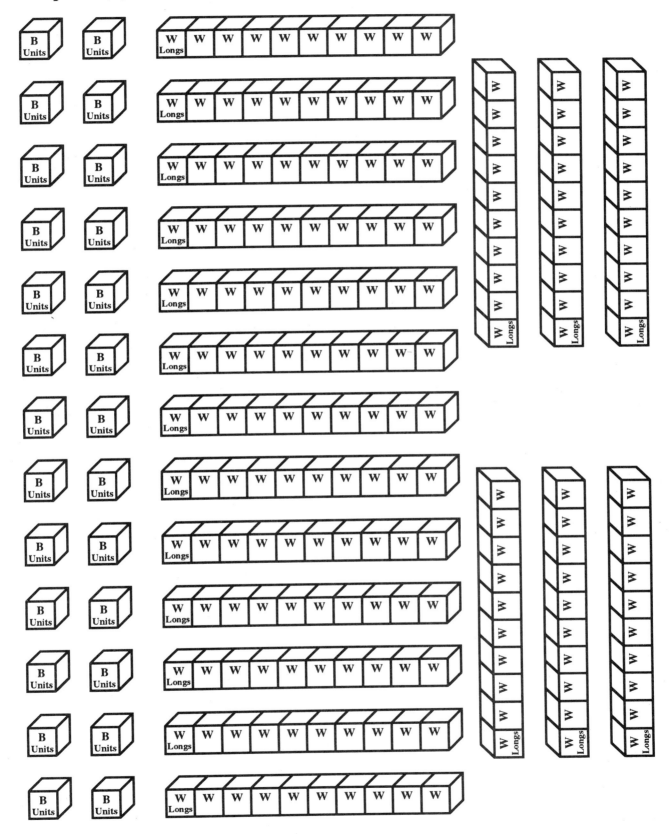

Chapters 2,3,7 – Base Ten Pieces

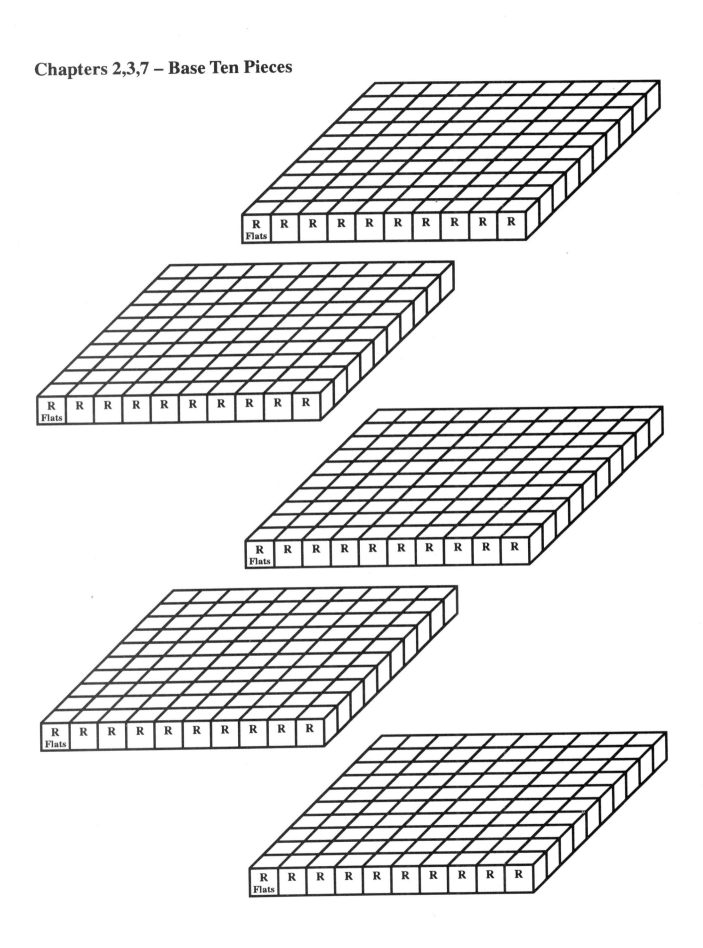

Chapters 2,3,7 – Base Ten Pieces

R Flats R R R R R R R R

R Flats R R R R R R R R

R Flats R R R R R R R R

R Flats R R R R R R R R

R Flats R R R R R R R R

Chapter 2 – Red, White, and Blue Chips

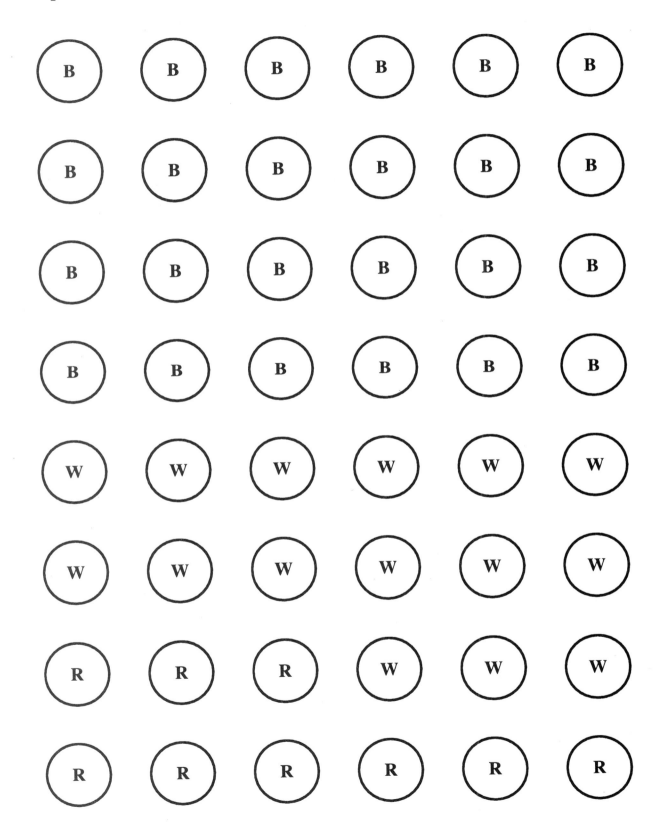

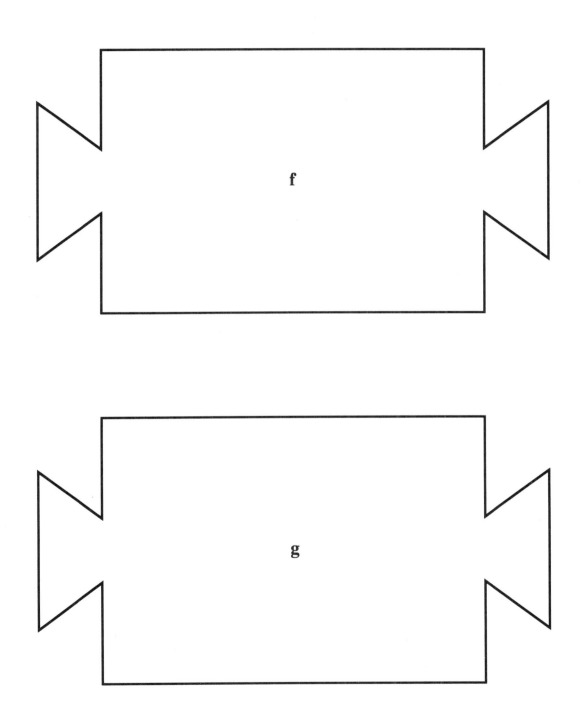

Chapter 4 – Clock 12 Card

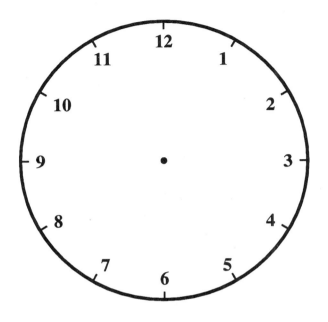

Chapter 4 – Modulo 3 Card

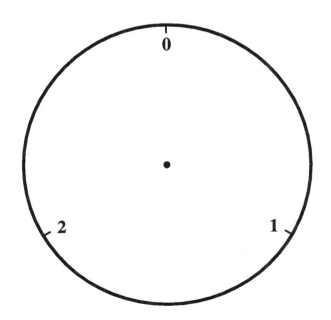

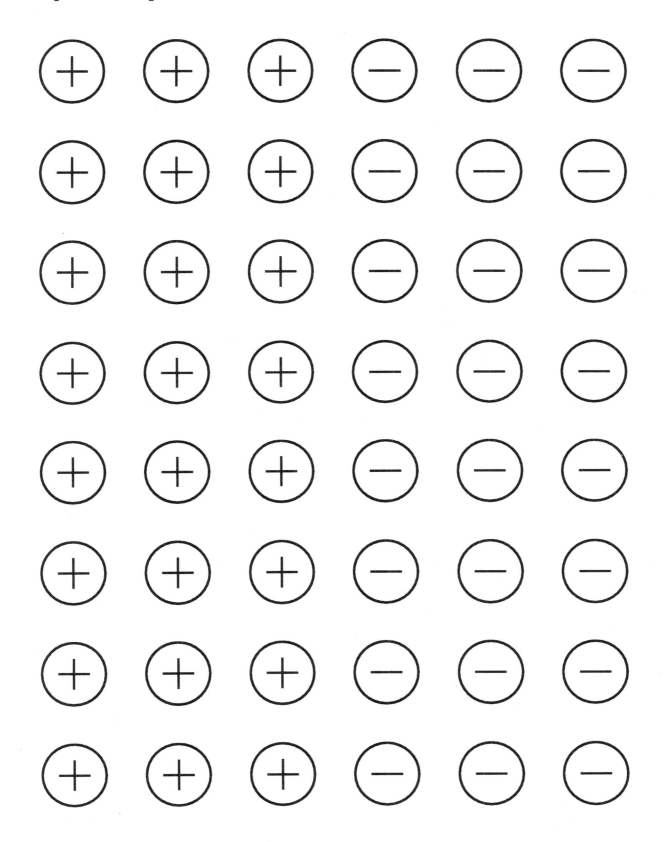

Chapter 5 – Box Diagram

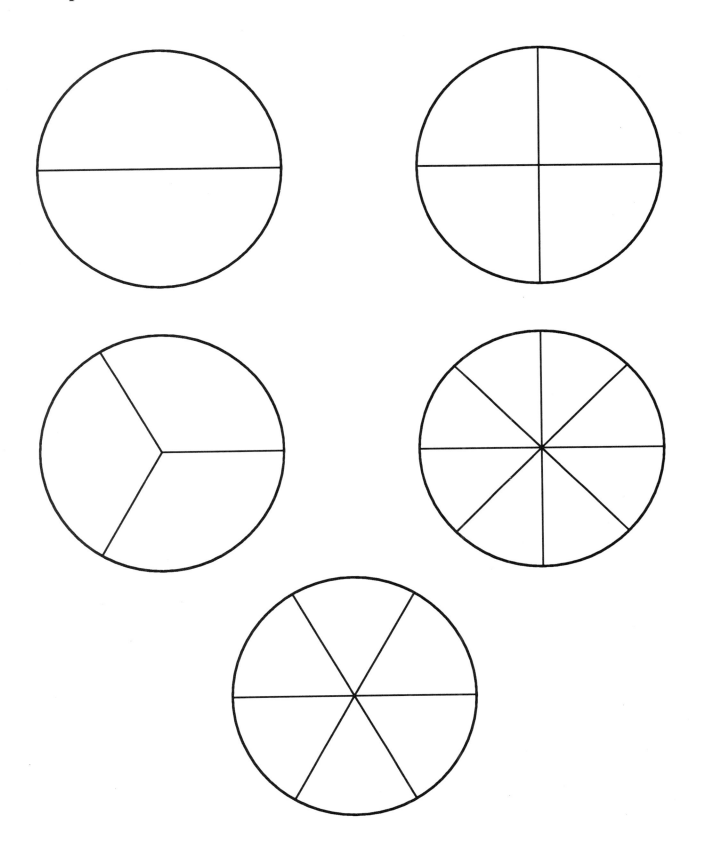

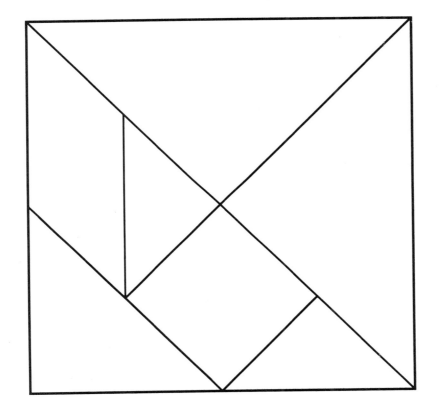

Chapter 8 – Metric Tape Measure

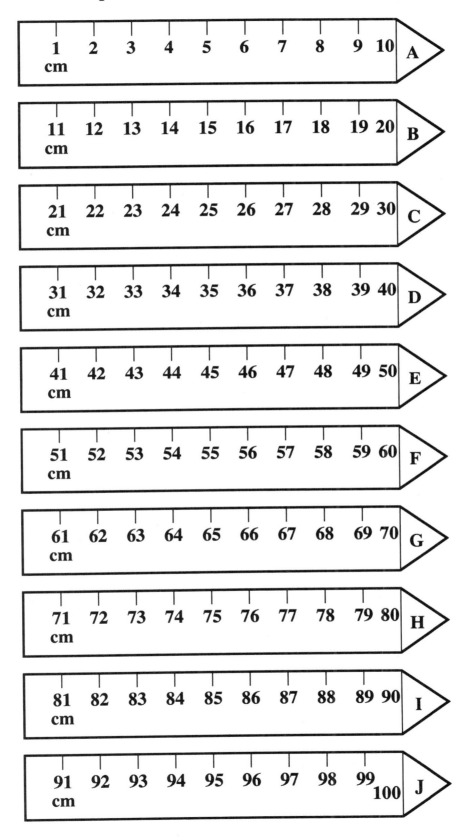

Chapter 10 – Polygon Cards #1

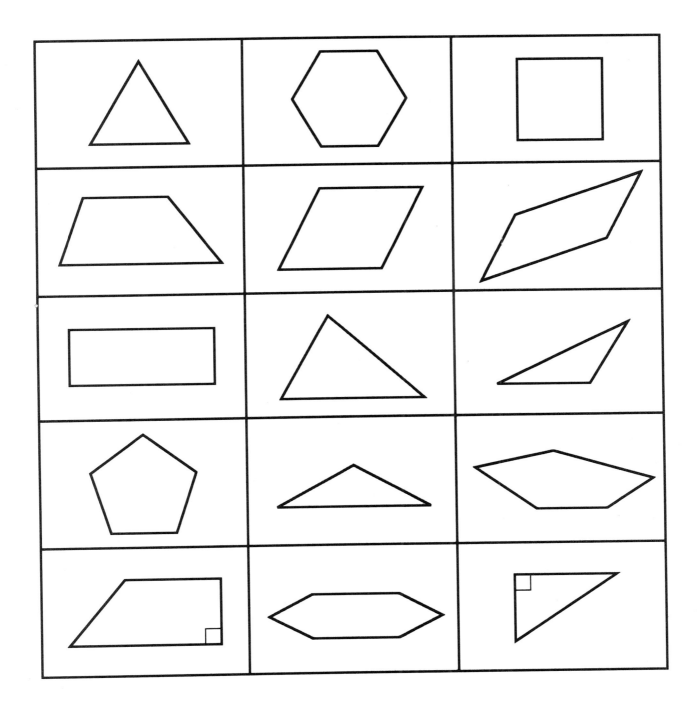

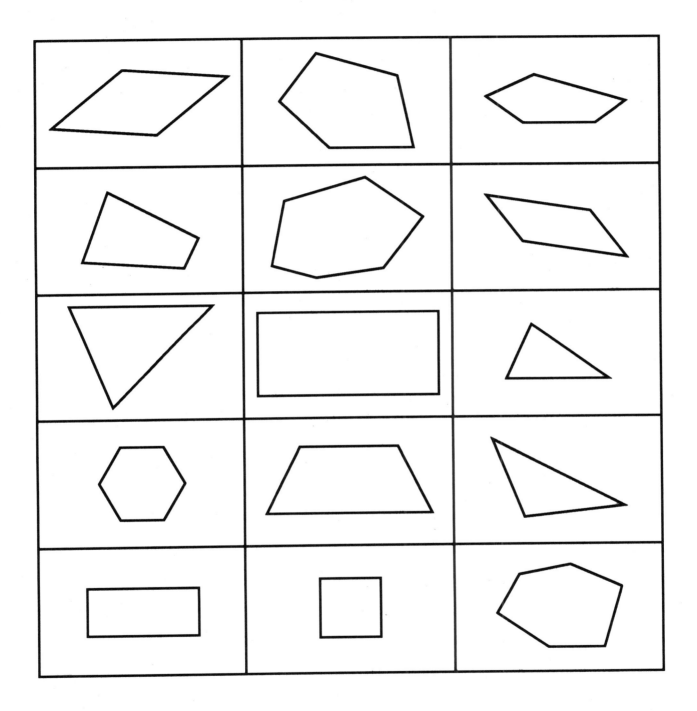

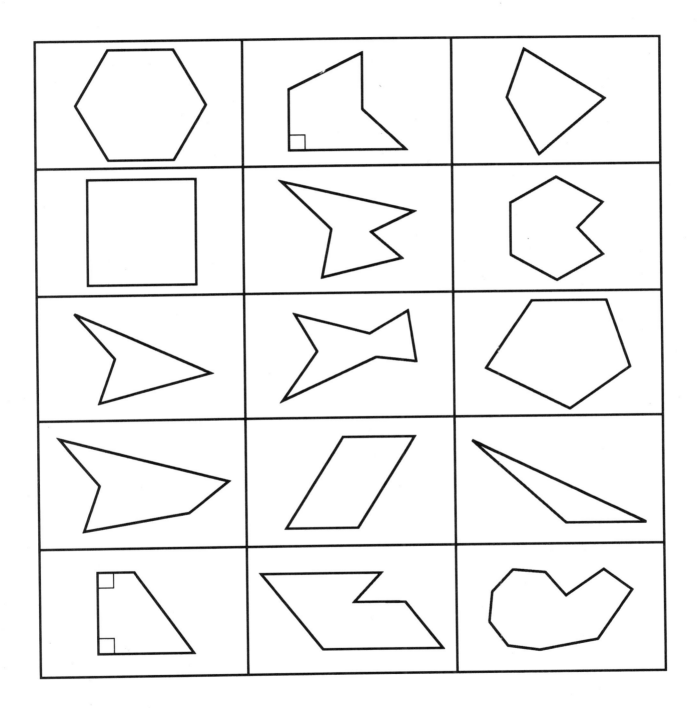

Chapter 11 – Metric Caliper

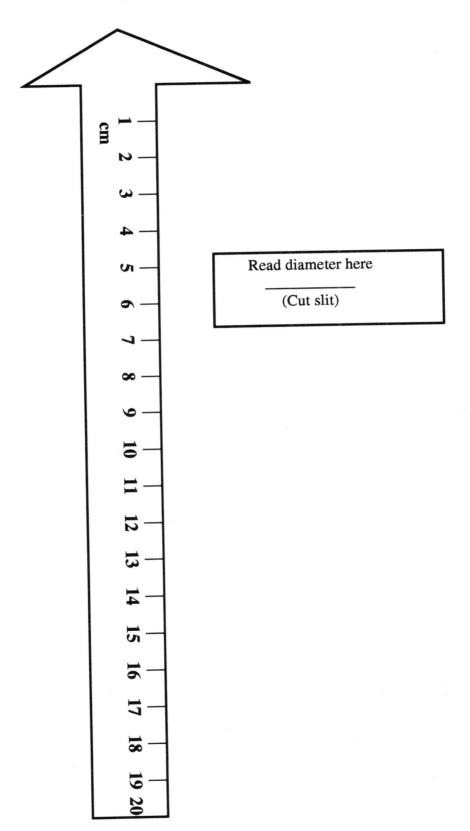

Read diameter here
_____
(Cut slit)

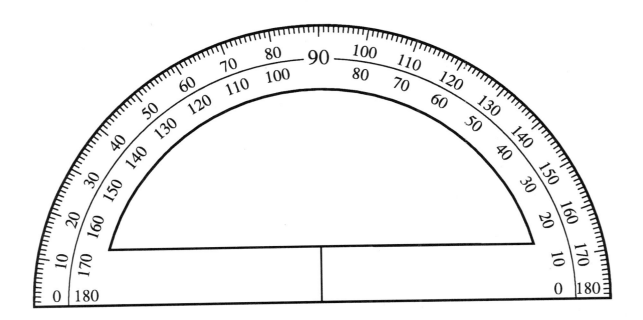

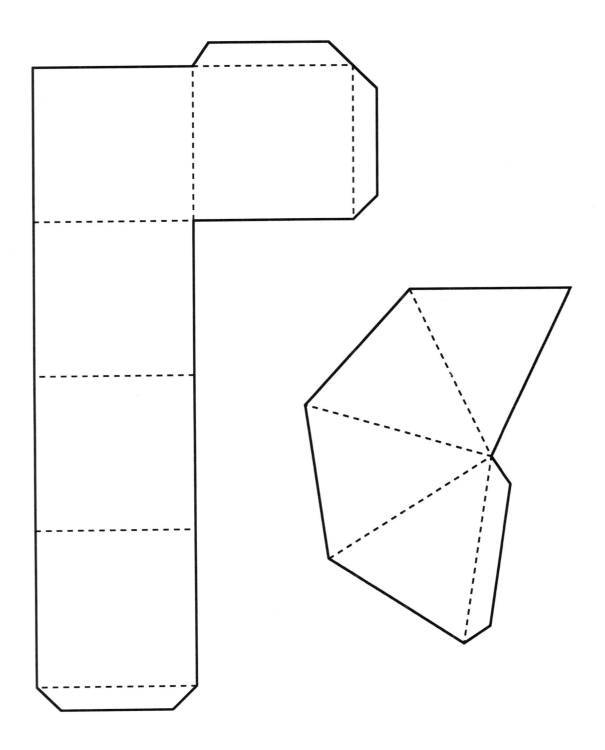

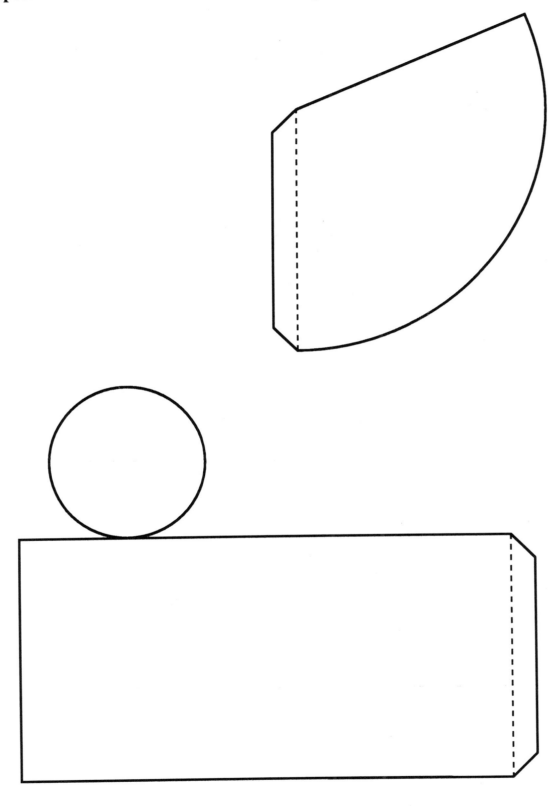

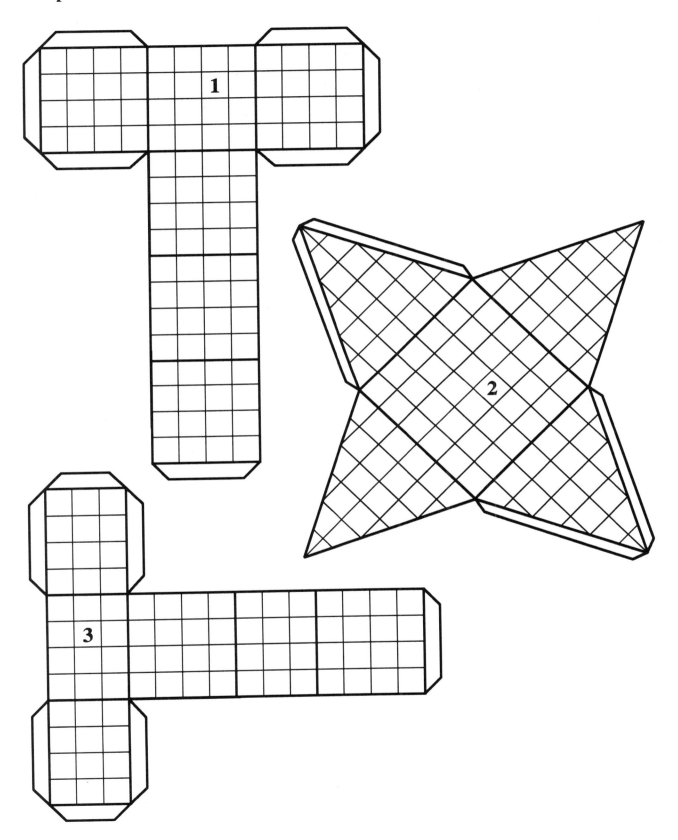

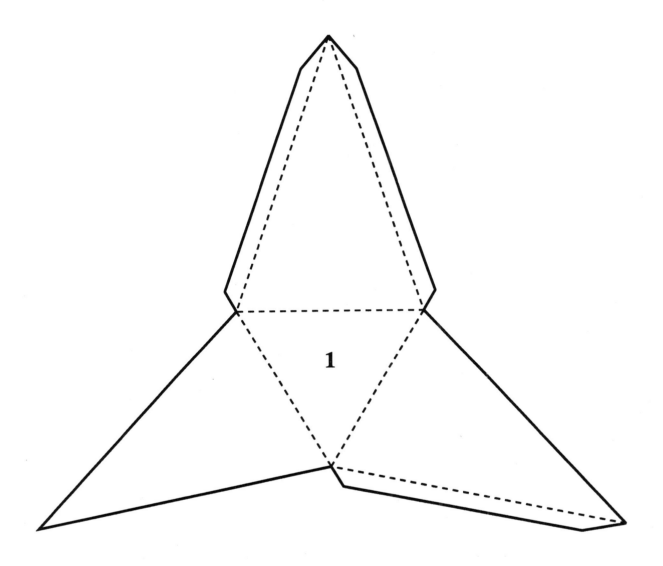

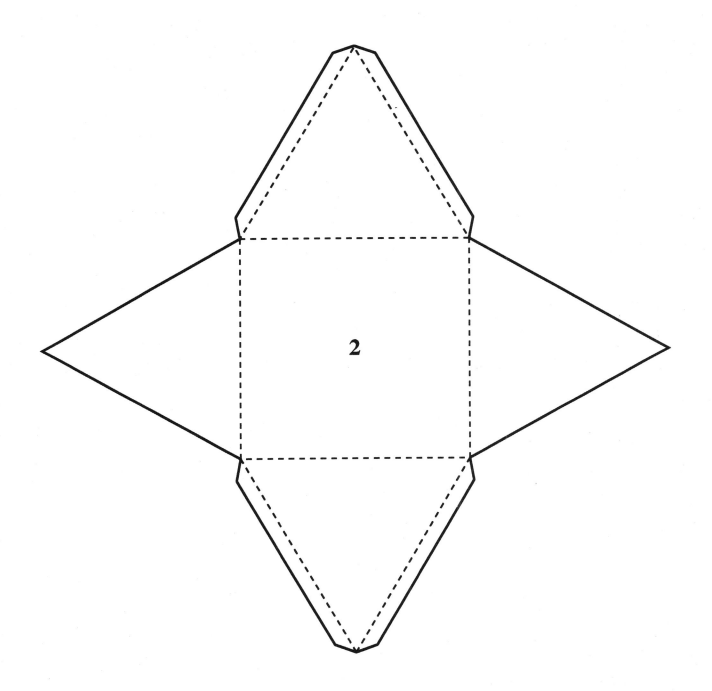

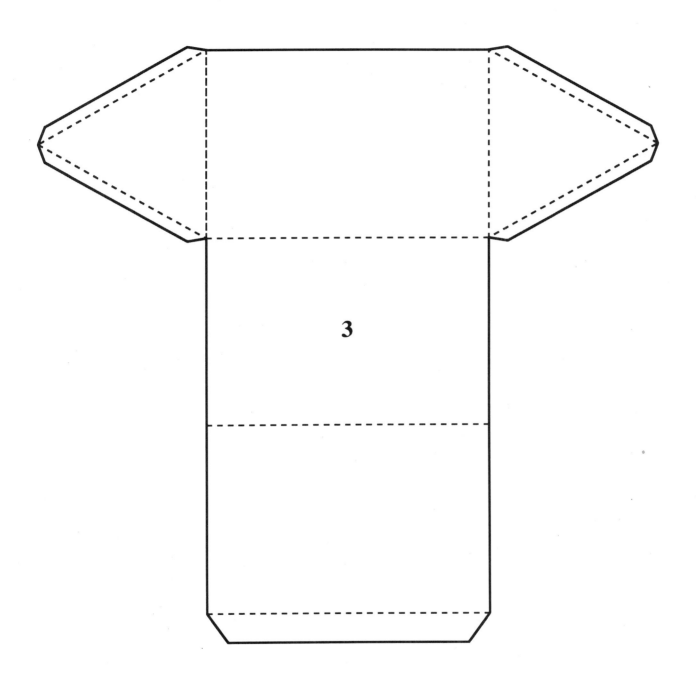

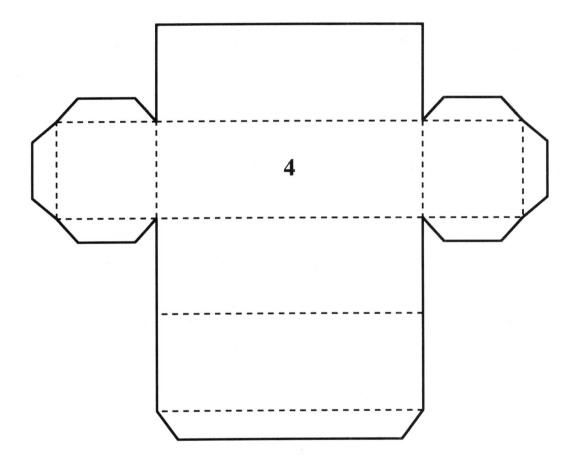

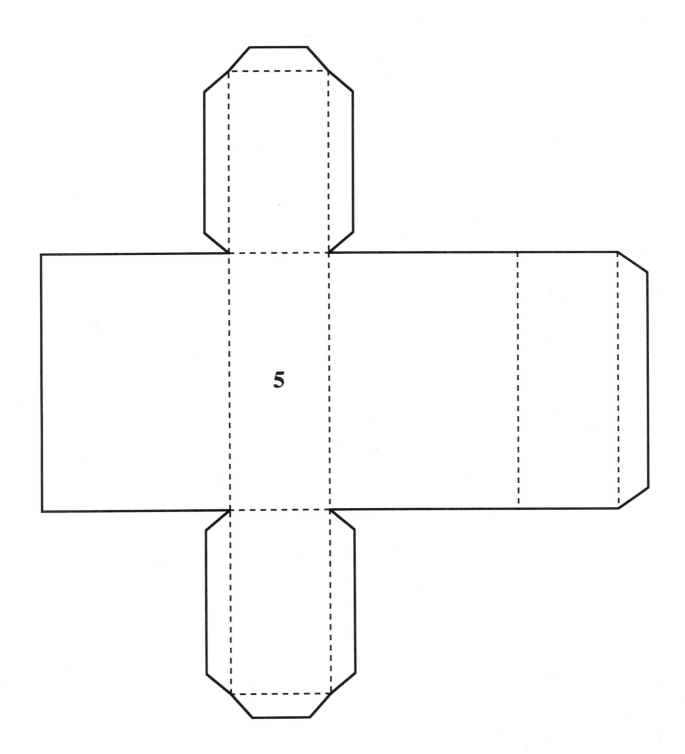

5

e :